Uncomfortable Ideas

Written By

BO BENNETT, PhD

Http://www.uncomfortable-ideas.com

eBookIt.com
365 Boston Post Road, #311
Sudbury, MA 01776

First printing - October 2016
Last Revised October 29, 2016

publishing@ebookit.com - http://www.ebookit.com

Copyright 2016, eBookIt.com
ISBN: 978-1-4566-2768-3

No part of this book may be reproduced in any form or by any electronic or mechanical means including information storage and retrieval systems, without permission in writing from the author. The only exception is by a reviewer, who may quote short excerpts in a review.

Table of Contents

Dedication i
Preface ii
    Cognitive Biases ii
    A Few Words About Me iv
    Political Correctness v
    The Structure of This Book vi
Part I: The Uncomfortable Idea 7
    What is an "Uncomfortable Idea?" 7
    Avoiding Uncomfortable Ideas 8
    Why it is Important to Entertain Uncomfortable Ideas and Accept Uncomfortable Facts and Truths 9
        Credibility..9
        Expose Dangerous Thinking...9
        Attempting to Solve the Wrong Problem 10
        Treating Symptoms and Not the Disease 11
        Understanding Unintended Consequences........................13
        Understanding Reduces Animosity....................................15
        Avoiding Manipulation 16
        The Importance of a Shared Reality 17
    Embracing Uncomfortable Ideas 17
    The Conscious, Unconscious, Group, and Individual Aspects of Avoidance 19
        Conscious, Group Avoidance 19
        Conscious, Individual Avoidance 22
        Unconscious, Group Avoidance 23
        Unconscious, Individual Avoidance 26

Part II: Uncomfortable Ideas and the Reasons Why We Avoid Them 28

Unconscious Avoidance 28

   Cognitive Dissonance..29

   Semmelweis Reflex 31

   Overcompensation 32

   Reaction Formation 35

   Intolerance of Nuance and Ambiguity..............................36

   Feeling Over Fact..42

   Uncomfortable and Unfalsifiable 44

   Protecting Sacred Beliefs..45

Conscious Avoidance...49

   Fear of the Slippery Slope 50

   Fear For Society 51

   We Don't Want To Be Seen As "Unpatriotic" 55

   The Desire to Hold Popular Views or the Fear of Social Response 61

   Fantasy is Sometimes Better Than Reality......................75

   The Work of Satan...77

   Fear of Entertaining Evil, Sick, or Immoral Thoughts 78

   Fear of Questioning / Refusal To Question Authority........81

   Fear of Confusing Support for Personal Desire 82

   Fear of Exposing Our Own Demons 84

Part III: Why We Refuse To Accept Uncomfortable Ideas 87

Evaluating Evidence 88

   Awareness 89

   Believability 89

   Comprehension..91

Belief-Related Cognitive Biases and Effects 92

    Backfire Effect *92*

    Belief Bias *93*

    Confirmation Bias *93*

    Ostrich Effect *94*

    Status Quo Bias *94*

Refusal to Accept Due to Refusal to Reject 94

Part IV: Some More Uncomfortable Ideas 96

The Self-Fulfilling Nature of Social Injustice 96

Love Isn't Always Beautiful, and You Don't Love Everyone 99

People Are Much More Selfish Than You Think 102

"Microaggressions" Are Less Common and Less Problematic Than People Think 104

Religious Ideas Are Protected By Motivated Reasoning More Than Any Other Class of Ideas 106

    Adam, Eve, and the 6000 Year Old Universe *108*

    The Soul *109*

    The Christian Bible *110*

    The "Goodness" of the Biblical God *111*

    Belief and Faith *114*

Being an Atheist Doesn't Make You Smarter and Certainly Not Better at Critical Thinking 114

    There is Evidence for God *115*

    No, Believing in God is Not the Same as Believing in Santa Clause. *115*

    Your Examples of History's Jesus-like Figures are Likely Made Up or Greatly Exaggerated *116*

    Evolution Does Not Answer the Question of Where we Came From *117*

    It is Foolish To Demand That Believers Prove That God Exists *117*

No, Theists Will Not Understand Why You Don't Believe In God When They Realize Why They Call Zeus a Myth 117

Everyone is Not Born an Atheist 117

Most Apparent Bible Contradictions Can Easily Be Explained 118

You Should Give President Trump or President Clinton Your Support 118

If You're Offended, You're Part of the Problem 120

Why We Choose To Be Offended 121

The Unconscious Factors That Influence Our Decision To Be Offended 122

The Person/Idea Distinction Myth 124

The Optimal Strategy 124

It is Okay to Change Your Mind 125

Concluding Thoughts 127

Part V: Uncomfortable Questions **129**

Life Partners 129

Love and Sex 130

Humanity 130

'Murica 131

Faith, God, and Religion 132

Metaphysics 134

Morality 135

Mental Health 137

Politically Incorrect 137

On the Lighter Side 139

Dedication

To all the people I've offended before
Who travelled in and out my door
My meaning they mistook
I dedicate this book
To all the people I've offended before

... and to Willie Nelson

Preface

"The suppression of uncomfortable ideas may be common in religion or in politics, but it is not the path to knowledge, and there's no place for it in the endeavor of science." - **Carl Sagan**

Imagine for a moment that a Neo-Nazi group is speaking at a local university. They are advertising that they are reaching out to the general public to help them understand that the Nazi party has been unjustly demonized, and they promise to discuss historical facts that will put the party in proper perspective. Do you go? Why or why not? Think about this for a moment. We'll refer to this question in the next section.

Cognitive Biases

I wrote this book for a general audience, but I don't shy away from technical terms—especially when they explain so nicely how we deal with, or **not** deal with, uncomfortable ideas. But I promise you this: when I do mention a technical term, I will do my best to explain it well and provide examples where appropriate.

Let's start with the term "cognitive bias." A *cognitive bias* is like an illusion for the mind. It is a deviation from rationality in judgment. Our brain did not evolve with rationality and reason as a goal; the only goals are reproduction and survival. Rationality is only needed to the extent that it supports one or both of those goals. Here's the big problem: evolution works over tens of thousands of years, and we have made dramatic changes to our social environment in the last several hundred years. Evolution hasn't had time to catch up. An example to which most us can relate, unfortunately, is overeating. We have a desire to overeat because food was scarce in our ancestral environment and the cost of starving was far greater then the cost of eating too much. Today, for most of us, there is no shortage of food, and we have a serious problem with obesity. The evolutionary trait that once aided in our survival is now killing us. Like the behavior of overeating, most cognitive biases are also relics of our ancestral environment that once helped us survive, but now, in the age of reason, are problematic.

Some just make us look silly, some lead to poor judgments and decision making, some threaten our lives, and some actually are responsible for killing us.

Consider *stereotyping*, which is a cognitive bias that allows us to quickly and efficiently (but not always accurately) make judgments about people. Tens of thousands of years ago, if our ancestors were approached by individuals who looked different from them, it was a safe bet to assume the strangers were a risk. They didn't have the luxury of time to get to know all about the strangers. If they hesitated in taking action, they might die. Today, stereotyping has become less effective (although far from useless) since there is less risk associated with taking the time to learn about others, and stereotyping is now more of a liability to us than it is an asset.

Which groups are more likely to avoid uncomfortable ideas? To answer this, we can turn to research in cognitive science that has focused on the question, "who tends to be more biased?" Republicans or Democrats? Christians or atheists? Men or women? The answer is a bit tricky since it depends on the bias being studied,[1] the passion the members have for their group,[2] and the metacognitive abilities of the members (the ability to think about their thinking process),[3] just to name a few of the factors. Perhaps the most important point in understanding biases is that the biases are not correlated with general cognitive ability,[4] that is, **intelligent people are not immune to biases**. Social scientist Keith Stanovich has done extensive research in

---

[1] Rudman, L. A., & Goodwin, S. A. (2004). Gender differences in automatic in-group bias: why do women like women more than men like men? *Journal of Personality and Social Psychology, 87*(4), 494–509. http://doi.org/10.1037/0022-3514.87.4.494

[2] Iyengar, S., & Hahn, K. S. (2009). Red Media, Blue Media: Evidence of Ideological Selectivity in Media Use. *Journal of Communication, 59*(1), 19–39. http://doi.org/10.1111/j.1460-2466.2008.01402.x

[3] The Importance of Cognitive Errors in Diagnosis and Strategies to Minimize Them : Academic Medicine. (n.d.). Retrieved from http://journals.lww.com/academicmedicine/Fulltext/2003/08000/The_Importance_of_Cognitive_Errors_in_Diagnosis.3.aspx

[4] Stanovich, K. E., & West, R. F. (2008). On the relative independence of thinking biases and cognitive ability. *Journal of Personality and Social Psychology, 94*(4), 672–695. http://doi.org/10.1037/0022-3514.94.4.672

the area of reasoning[5] and proposed that one's ability to reason effectively, that is to recognize and avoid biases largely responsible for our avoidance of uncomfortable ideas, is a separate intelligence just like emotional intelligence differs from general intelligence. Rational intelligence is an intelligence that is learnable. This book will help you become more intelligent in the area of rationality primarily by helping you to learn and recognize the biases that work against this intelligence. This improves your *metacognition*—your ability to think about how you reason.

A Few Words About Me

As a social psychologist, my goal is to see issues as objectively as possible while recognizing my own biases. For full disclosure, I am a white, cisgender, heterosexual, married, well-educated, upper-middle class male. I don't have strong political beliefs, but I am definitely left of center. I am an atheist with a naturalistic worldview, but I can certainly appreciate religions for the benefits they offer some people and communities. Given my background, I cannot speak to the lived experiences of the members of the transgender and gay communities, non-whites or women, but I can explore related topics scientifically, objectively, and without passion or ideology. If we want to know about climate change, we're better off getting our information from climatologists than from Eskimos, even though Eskimos experience the effects of climate change. Knowledge and experience are not the same.

I've done my best to being fully objective in creating this book. This has allowed me to present some uncomfortable ideas that I don't necessarily agree with but know that other people do. I am not presenting a balanced assessment of the ideas because virtually all of us have heard the "arguments" against these ideas already. I am presenting arguments for ideas that you likely have not heard before. Just because this book is not balanced, it doesn't mean it is not fair or that the arguments are not strong and factual. I have cited all claims where data support the claims, and when I come to my own

---

[5] Research on Reasoning. (n.d.). Retrieved from http://keithstanovich.com/Site/Research_on_Reasoning.html

conclusions I have done my best to reasonably justify those conclusions.

The goal of this book is to explore many uncomfortable ideas that are often not expressed, entertained, or accepted for a myriad of reasons. If I did my job right, you will no doubt be offended or at the very least be made uncomfortable by many of these ideas. Based on the ideas I do support, you might call me a bigot, racist, misogynist, snob, elitist, sympathizer, shill, godless heathen, or perhaps just an asshole. With the exception of the "godless heathen" label, I don't think I am any of those, but I also think my jeans from high school still fit me fine.

Political Correctness

Political correctness is defined as "the avoidance, often considered as taken to extremes, of forms of expression or action that are perceived to exclude, marginalize, or insult groups of people who are socially disadvantaged or discriminated against." Think of social behavior on a continuum. At one end, we have overtly racist, sexist, and mean-spirited forms of expression or action directed towards those who are incapable of defending themselves due to lack of power. On the other end of the continuum, we have **any form of expression or action** that is **interpreted** as overtly racist, sexist, and mean-spirited. Political correctness exists between the two extremes. This means a socially unaware person can think she is politically correct by referring to a black person as a "negro" (and avoiding the other N-word) where most others would consider her comment politically incorrect. Conversely, a college student could start a protest over the term "Black Friday" connecting the day after Thanksgiving somehow to slavery, then call it "racist." There is no universally agreed upon ideal level of political correctness and what is extreme to one person might be perfectly reasonable to another. Be prepared to argue for your opinion and convince others why you are right.

Uncomfortable Idea: You are not the authority or standard on what is or is not politically correct. You don't have moral superiority; you have the illusion of it.

The Structure of This Book

In part one, we look at the meaning of "uncomfortable idea," specifically what uncomfortable ideas are, what it means to avoid them, and why it's so important to entertain them and, at times, embrace them.

Part two deals with the most common unconscious and conscious reasons why we avoid uncomfortable ideas and includes dozens of examples, most of which will fall outside your comfort zone.

Part three looks at why we refuse to accept uncomfortable ideas that we would likely accept if they weren't uncomfortable.

In part four, you are presented with several uncomfortable ideas that should make you rethink many of your core beliefs.

Finally, in part five, you will find a listing of over a hundred uncomfortable questions that will make excellent discussion questions for college classes, social media, or fun questions to break out at a party—assuming you don't mind some heated discussions.

Not everyone will find all of these ideas uncomfortable, but the chances are most of you will find most of these ideas uncomfortable. Don't avoid them; entertain them and either accept them or educate yourself as to why they shouldn't be accepted, so you will be prepared with reasons as to why the ideas are bad when someone is attempting to convince you otherwise. This is the foundation of critical thinking.

Part I: The Uncomfortable Idea

What is an "Uncomfortable Idea?"

Simply defined, an *uncomfortable idea* is an idea that makes you uncomfortable. This is a *subjective concept* meaning that any given idea can be uncomfortable to you but not to another person or vice versa. More specifically, **an uncomfortable idea is a thought that is difficult to entertain due to real or anticipated psychological pain or social consequences that result from entertaining the thought.**

Recall the opening question about attending the Neo-Nazi event. For most of us, just considering attending this event makes us feel uncomfortable, and we didn't even get to the ideas presented in the event. Perhaps you wouldn't attend the Neo-Nazi event simply because you have no interest whatsoever in the topic. You don't care if Hitler had a good side or if there were aspects of the Nazi party that made good social and economic sense. It wouldn't surprise you, offend you, or make you uncomfortable in any way—you just don't care. There are countless ideas and even more perspectives on those ideas. One couldn't possibly entertain them all in a lifetime let alone someone who works, has a family, and enjoys one's spare time. However, any ideas that fit into this category would, by definition, not qualify as "uncomfortable ideas." Perhaps if more people supported the "nice guy Hitler" idea, then it would become an uncomfortable idea worthy of being entertained, but for now, it is simply an idea unworthy of our consideration. It is up to those who are trying to get us to consider the idea to convince us as to why it matters.

In addition to being apathetic to the idea (i.e., not giving a rat's ass), you might be well informed and not likely to be exposed to any new information that will change your mind. Perhaps you have studied Nazi history and are well aware of the arguments and reasons presented by Nazi sympathizers. Your "avoidance" of the idea would more accurately be a refusal to waste your time on hearing information you already know, or can reasonably anticipate hearing. Indifference and being well-informed aside, ideas are often avoided for a reason.

Avoiding Uncomfortable Ideas

We avoid uncomfortable ideas in three main ways: we avoid *entertaining* them, we avoid *accepting* them, and we avoid *expressing* them. These processes can be deliberate or done subconsciously or have components of both. Many of the same reasons we avoid entertaining uncomfortable ideas apply to why we avoid accepting and expressing these ideas. Refusing to entertain an uncomfortable idea is a conscious decision not to think about, investigate, or consider evidence for the idea. There are dozens of reasons why we do this. Many times there are multiple reasons combined that cannot be articulated, but we just "know" that an idea is not up for debate or consideration. The problem is, virtually all of these reasons are irrational; based on biases, cognitive effects, heuristics, fallacies; or other obstacles in reason.

We can, and do, accept ideas without entertaining them. We do this all the time when we trust authority, when we are raised with a certain idea, when we are cognitively exhausted, or if we are gullible and just not very good at critical thinking. While accepting a good idea for a bad reason is better than accepting a bad idea for a bad reason, it's best to accept a good idea for a good reason. In other words, entertaining the ideas we do accept or thinking critically about them is an important component of reason.

Back to our opening question. If you were to immediately reject the invitation to attend the presentation by the Neo-Nazi group simply because you think Neo-Nazis are "animals," you would be refusing to entertain the ideas in what might be a mostly subconscious process. If you were to agree to go but sat through the entire event with your arms crossed uncritically dismissing every point that was made, you would be refusing to entertain the ideas in what would most likely be a deliberate thought process where you decided ahead of time that if you heard anything that made sense, it would only be propaganda and lies. Perhaps you did entertain the ideas critically while recognizing your biases, and now comes the time when you decide if you accept the ideas or not. This is the most difficult part. As we will explore in this book, you might have perfectly reasonable justifications for not accepting the idea, but there are many ways in which our brain "protects us" against uncomfortable ideas, no matter how factual and true they might be. Remember, truth-seeking and understanding reality are not the goals of evolution; survival and reproduction are.

Why it is Important to Entertain Uncomfortable Ideas and Accept Uncomfortable Facts and Truths

So what if you choose not to listen to a Neo-Nazi justify his ideas? So what if you refuse to entertain ideas that contradict with your religious beliefs? So what if you support your political party's ideas 100% no matter what? Depending on the idea, the costs can range from embarrassing to catastrophic.

Credibility

To some people, credibility still matters. If you are arguing for one side of an issue and you fail to acknowledge valid points by your opponent, or worse, outright reject valid points, you will lose any credibility you do have with your opponent and you are likely to lose credibility with the audience, as well. I recently heard a debate where one of the debaters was defending consequentialism as a moral theory. His opponent made the point that we rarely know the long-term consequences of our actions and most people are lousy at predicting even short-term human behavior in response to an action. Rather than accept this uncomfortable idea as a valid point (which it is), the debater defending consequentialism refused to address this criticism. Perhaps it was just debate tactic and he was counting on avoiding the criticism being a better strategy than acknowledging it. But the whole exchange did result in me concluding that the debater defending consequentialism really didn't think his position through very well. His credibility was lost with me due to his failure to acknowledge an idea that was clearly uncomfortable to him.

Expose Dangerous Thinking

The world is full of some really dangerous people with even more dangerous ideas. When we put restrictions on the expression of ideas, we make it more difficult to identify potentially dangerous people and their dangerous ideas. While it might be more comfortable avoiding these people and their ideas, and pretending they don't exist, the better strategy is confronting them and doing our best to explain why we think their ideas are dangerous. Isolation breeds extremism; integration promotes moderation.

The casual use of deliberately insulting labels such "racist," "misogynist," "homophobe," and "xenophobe" might feel satisfying and might even be perfectly justifiable, but rarely causes the accused to change their way of thinking. Instead, it lets people with dangerous ideas know that their ideas are socially unacceptable, which is not the same as being wrong or dangerous. We need to encourage the free exchange of ideas, not banish those with dangerous ideas into exile where their ideas fester and eventually are expressed as behaviors and possibly even devastating actions.

Attempting to Solve the Wrong Problem

If you're convinced that the United States has a problem with racist cops, then you would focus on the problem of racism—no doubt a worthy social issue. But what if you consider the less socially acceptable ideas that the United States has a problem with police brutality and use of excessive force? Entertaining this idea might lead you to facts and data that justify that conclusion (such as the percentage of white suspects who are also unjustly shot), which would turn your attention to what might be the larger problem. Racism might have an effect on the number of black suspects being unjustly gunned down by cops, but excessive force used by police and the laws that protect such force might have a much greater effect on the number of all suspects, regardless of skin color, being unjustly gunned down by cops. If we don't entertain the alternative ideas, even if they go against our ideology, personal experience, or anecdotal evidence, we can be wasting our time on solving the wrong problem.

Uncomfortable Idea: The police in the United States have a much greater problem with excessive force than racism.

Here is a fact. The U.S. locks people up at a higher rate than any other country.[6] Clearly, the United States has a problem with crime. Or does it? What if we took a closer look at the laws, policies, and social norms and questioned them? Should we be locking people up for using recreational drugs? Should we be locking people up for selling and buying sex? Should an 18-year-old girl serve jail time for having sex with her 17-year-old boyfriend? The United States has many laws that are based on outdated ideology, iron-age morality, and incorrect assumptions. Unless we deal with the uncomfortable idea that our concepts of morality and justice might be way off, we will continue to create more crime by making more things illegal. In other words, our outrageous incarceration rate in the "land of the free" is a symptom of the the real problem. The problem, however, is not an excessive number of evil people hurting other people; it is locking people up who don't live up to our moral standards.

Uncomfortable Idea: We are locking too many people up because of our rigid understanding of morality and justice.

Treating Symptoms and Not the Disease

When we have a cold we eat chicken soup, take extra vitamin C, and have our chakras realigned while placing healing crystals in our navel. This is all fine and dandy if it makes us feel better, but it does nothing to make the cold go away. When it comes to the common cold, the best we can do with our current medical knowledge is treat the symptoms but not the disease. If we could treat the disease, this would

---

[6] Yes, U.S. locks people up at a higher rate than any other country. (n.d.). Retrieved from https://www.washingtonpost.com/news/fact-checker/wp/2015/07/07/yes-u-s-locks-people-up-at-a-higher-rate-than-any-other-country/

clearly be the preferable option that would prevent a lot of future suffering. When it comes to social issues, our desire to avoid uncomfortable ideas can cause us to focus on the symptoms while ignoring the disease.

Why is saying "people of color" appropriate but saying "colored people" is offensive to most people in 2016? Is it the word "of" that has some special power to diffuse racism? Does putting the adjective before the noun make it offensive? The reasons why terms become offensive and sometimes become acceptable after being deemed offensive are mostly due to association and negative connotations. There are terms used to label every race, class, nationality, sexual preference, gender identity, physical and mental difference, and what's acceptable is constantly changing. While we are all so concerned about terminology, the underlying problems of prejudice and discrimination continue. The negative connotations that become associated with the labels are just a symptom of the prejudice and discrimination.

Uncomfortable Idea: No matter how many times marginalized groups change their preferred label, they will still be marginalized unless the real difficult underlying problems are addressed.

In our age of political correctness, we might be infuriated by the expression of any negative stereotype, but some stereotypes are supported by data. If we dismiss the stereotype as a form of prejudice and focus on suppressing the stereotype, the underlying problem that gave rise to the justified stereotype and associated prejudice will continue. Consider that black, adult, males are seen as more violent than white adult males[7]—this is a common stereotype. We can blame this on racism and try to get people to drop this stereotype, but that is

---

[7] Duncan, B. L. (1976). Differential social perception and attribution of intergroup violence: Testing the lower limits of stereotyping of Blacks. *Journal of Personality and Social Psychology, 34*(4), 590–598. http://doi.org/10.1037/0022-3514.34.4.590

difficult to do when the stereotype is supported by the data.[8] What we need to do is look at why blacks are more likely to commit a violent crime (or at least found guilty of it—yes, there is no question that due to prejudice, blacks are more likely to be found guilty[9]). Once we solve the problem, it will be reflected in the data, and the stereotype will become an unjustified one that will weaken over time and likely disappear. Unless the underlying causes are addressed, there is little chance the symptoms (stereotypes) will go away.

Uncomfortable Idea: Stereotypes are sometimes supported by the data and reflections of reality.

Understanding Unintended Consequences

When we get caught up in ideology and political correctness, we overlook the downside of our actions. Yes, there are downsides to just about every action even if one pretends there are not. I spoke at a conference for skeptics recently that adopted a "color communication badge" policy originally created for those on the autism spectrum. In short, conference attendees had the option of placing a sticker on their badge that indicated the level of social interaction they wanted. For example, according to the posted policy:

> *Showing a red square sticker means that the person probably does not want to talk to anyone, or only wants to talk to a few people. The person might approach others to talk, and that is okay; the approached people are welcome to talk back to them in that case. But unless you have been told already by the badge-wearer that you are on their "blue list," you should not approach them to talk.*

---

[8] Expanded Homicide Data Table 6. (n.d.). Retrieved from https://ucr.fbi.gov/crime-in-the-u.s/2013/crime-in-the-u.s.-2013/offenses-known-to-law-enforcement/expanded-homicide/expanded_homicide_data_table_6_murder_race_and_sex_of_vicitm_by_race_and_sex_of_offender_2013.xls

[9] T. U. of P. L. S. S., Philadelphia, & map215.898.7483, P. 19104. (n.d.). New study by Prof. David Abrams and co-authors confirms racial bias in criminal sentencing. Retrieved from https://www.law.upenn.edu/live/news/2170-new-study-by-professor-david-s-abrams-confirms

The reason for implementing this kind of policy at a convention where the perception of those on the autism spectrum is unlikely to be much different than the general population, is stated in the same policy:

Color communication badges also help all people, abled or disabled, to more easily and effectively let people know whether they want to be approached for conversations or not. This can create a positive impact on the social atmosphere where communication badges are being used.

While this policy sounds great for those on the autism spectrum, there is a significant downside that, from a social psychological perspective, I would argue is more damaging than helpful. Social interaction is not easy for everyone, but "difficult" has never been a good reason not to do something or persist at improving. Communication is like a muscle; if we don't use it, we lose it. We might try to read people and get it completely wrong, but this is how we learn. We might have a difficult time communicating to others in a non-awkward way that we don't want to be talking to them, but with each interaction, we get better at it—unless we avoid all potentially uncomfortable interactions through "communication badges."

Electric mobility scooters make it easier and more effective for able-bodied people to shop in the mall, but most of us can agree this is not a good idea if we care about maintaining our ability to get around unassisted (think of the Disney movie *Wall-E*). Likewise, unless the entire world agrees to implement communication badges, we will lose our ability to easily and effectively interact with others who don't use this badge system. Refusing to implement an "accessibility" policy for any reason is generally not an idea that's very popular. Defenders of such policies think they have the moral high ground by protecting the disabled, but don't realize that they are also creating the disabled.

Uncomfortable Idea: In our efforts to be accommodating we can do more harm than good to those we are trying to help. There is a fine line between accommodation and coddling that we often cross when attempting to be politically correct.

Another example of the unintended consequences of avoiding uncomfortable ideas has to do with *identity politics*, or a political style that focuses on the issues relevant to various groups defined by a wide variety of shared personal characteristics. Some of the more common characteristics include race, religion, sex, gender, ethnicity, ideology, nationality, sexual orientation, gender expression, culture, shared history, and medical conditions. We celebrate marginalized groups that focus on promoting equality for their group, but we don't like the idea that this kind of deliberate self-classification based on superficial traits could have some significant unintended consequences. While it may feel cathartic to commiserate with a group of people just like you while demonizing those who are not like you, focusing on our whiteness, blackness, maleness, femaleness, gayness, straightness, or any other "ness" robs people of their individualism and is the polar opposite of the proven strategies that have been used to reduce prejudice and bring groups together. These proven strategies include focusing on similarities and common objectives.

Uncomfortable Idea: Identity politics is a dangerous game where the unintended consequences could be even greater prejudice against the group.

Understanding Reduces Animosity

Depending on how passionate one might be about certain issues, one can dislike or even **hate** people who hold opposite views on those issues. If we want to understand why someone holds the idea they do, we need to entertain the idea. What we often realize is that biological differences, different life experiences, or different values account for these different ideas. In the case of biological difference, the fairly new area of neuroscience shows us that biological differences in the brain affect how we process and understand information including political and religious beliefs. Different life experiences may include the indoctrination we received as a child, education or lack thereof, or some strong emotional experience that had a great impact on how we see the world. Even though we all may claim that we value the same things, how we define those things and to what degree we value each thing varies greatly. For example, two people both might value justice and compassion, but one might support the death penalty because to them, death is a just punishment for murder and justice is more important to them than compassion. Or perhaps, one might claim that they value compassion more, but it is through their sense of compassion for the loved ones of the victim that they support the death penalty. Once we understand why people hold the ideas they do, we are far less likely to hate them for it.

Uncomfortable Idea: Opposite views can often be reduced to prioritizing different values, where there is no wrong or right.

Avoiding Manipulation

Very often, one who presents fringe ideas is well aware of the common objections to the idea and like a good salesperson has crafted answers that address the objections. These answers can be fallacious, flawed, or outright lies designed to get one to buy into the idea. Once a person has accepted the idea, evidence against the idea has less of an effect on them rejecting the idea. This is why it is important to entertain an idea without pressure and have enough time to evaluate the arguments for and against the idea critically. For example, if someone tried to sell you on the idea that the earth was flat, they might claim that the horizon always rises to meet eye level, which is *impossible* on a ball earth. Not having investigated this, you can be skeptical, but you would really not be able to refute their claim. A few minutes of research, however, would demonstrate that this claim is simply untrue. The point is not to debunk what you have predetermined to be a false claim (this is reactionary thinking, not critical thinking), but to entertain the claims and evaluate them for accuracy.

The Importance of a Shared Reality

One unifying force of humanity is our shared reality. Similarities bring people together while differences tend to tear us apart. We are able to thrive as a species because we are a *social species*, one that is capable of creating a system in which each person's unique contributions benefit the group. For example, a farmer will raise cows that provide milk, a doctor will help people stay healthy, and a builder will build homes. This kind of system works because we share a common reality where people need food, healthcare, and shelter. This reality is founded on reason, logic, evidence, and experience. But this system breaks down when reason, logic, facts, and evidence are discarded, and experiences are interpreted in heavily biased ways. A group of people who feel that their god wants them to kill those who don't believe in their god, a group of people who reject facts of science and hinder the kind of scientific advancement that saves lives because it is inconsistent with their beliefs, or a group of political extremists who have been manipulated by emotional arguments, are examples of how personal realities contribute to the suffering of a society. The starting point of cooperation is a shared reality.

Uncomfortable Idea: In order to live harmoniously with others, when your personal beliefs are in conflict with our shared reality that is based on reason, logic, facts, and evidence, shared reality must take precedence.

Embracing Uncomfortable Ideas

There is a common misconception that we need to choose between happiness and some of the more "depressing" aspects of reality that are commonly seen as uncomfortable ideas. While I can think of a few cases where this might be the case, humanity is far more resilient than we give ourselves credit for. A landmark study published in 1978 demonstrated that lottery winners are no happier and paralyzed accident victims are no less happy a few months after their life-changing events. In addition, the lottery winners were often less happy than they were prior to winning the lottery because they took less pleasure in mundane events.[10] Unlike being paralyzed, accepting uncomfortable ideas can be relatively benign such as realizing you're not as good looking as you think. However, they can also be even more devastating, such as realizing that there is no benevolent god looking out for your well-being after spending a lifetime as a devout Christian. Optimists undoubtedly do better embracing uncomfortable ideas because these ideas often involve perspective. To give you a personal example, I spent the first 35 years of my life believing in an afterlife—the comfortable idea that I was going to live for eternity. Then I had to go ahead and start studying philosophy, world religions, and psychology. And if that weren't enough to turn my worldview upside down, I read the Bible from cover to cover—not just the warm and fuzzy parts. I could no longer believe in an afterlife, and that was difficult for me to handle, but only at first. I am an optimist, and I quickly began to realize that every moment I am alive is now more precious. I don't have eternity to do things or enjoy time with my family; I just have this life. In the last nine years since that realization I started several business, wrote and published seven books including my memoirs, earned a master's degree in generally psychology, earned a PhD in social psychology, lost 30 pounds (and kept it off), vacationed in over a dozen countries, accepted a teaching position at a local college, spent a lot of quality time with my wife and kids, and I am currently crossing off the last item on my bucket list—writing a screenplay. Just because an idea is uncomfortable at first, does not mean it will remain uncomfortable.

---

[10] Brickman, P., Coates, D., & Janoff-Bulman, R. (1978). Lottery winners and accident victims: Is happiness relative? *Journal of Personality and Social Psychology, 36*(8), 917–927. http://doi.org/10.1037/0022-3514.36.8.917

The Conscious, Unconscious, Group, and Individual Aspects of Avoidance

Within the context of avoiding uncomfortable ideas, "avoiding" can refer to a) keeping the idea from entering one's own thoughts or b) the conscious decision to not think about, investigate, or consider evidence for the idea. We avoid uncomfortable ideas consciously and unconsciously, as groups and as individuals.

Conscious, Group Avoidance

Conscious, group avoidance occurs when two or more people deliberately plan to keep themselves and/or others from exposure to or the entertaining of uncomfortable ideas. This kind of avoidance is common with universities, student groups, parents, and activist groups.

According to the Foundation for Individual Rights in Education (FIRE), since 2000, 82 total invited speakers were unable to speak at a university event because they were either formally disinvited, they voluntarily withdrew, or they were prevented from speaking due to substantial disruption by protesters in the audience.[11] There are many more speakers who are protested by student or faculty groups who still end up speaking despite the efforts to stop them. Take Michael Bloomberg for example. In 2014, when word got out that he was scheduled as the Harvard commencement speaker, several students protested because of Bloomberg's policies that they felt discriminated against minorities.[12] This is a reasonable concern and certainly not unjustified. The problem is, actions, behaviors, and policies are very often a reflection of a person's beliefs, moral code, and politics. Avoiding exposure to ideas from people whose actions, behavior, or policies we don't agree with is the same as avoiding them for their ideas that we find uncomfortable. By avoiding these ideas, we are creating an echo chamber environment where principles in group psychology such as *groupthink, group polarization,* and *memory*

---

[11] Disinvitation Report 2014 Infographic. (2014, May 28). Retrieved from https://www.thefire.org/disinvitation-report-2014-infographic/

[12] Students Question Selection of Bloomberg as Commencement Speaker | News | The Harvard Crimson. (n.d.). Retrieved from http://www.thecrimson.com/article/2014/3/11/students-question-bloomberg-speaker/

biases all but assure that ideas uncomfortable to the group become even more uncomfortable to that group. For example, *group polarization* is the phenomenon that when placed in group situations such as student groups or entire student populations, people will form more extreme opinions than when they are in individual situations.

Uncomfortable Idea: Refusing to allow people to share their ideas, no matter how dangerous you may think their ideas are, can often do more harm than good.

Parents are well known from shielding their children from uncomfortable ideas. This is called *parenting*. Although there are as many philosophies to parenting as there are parents, virtually all parents would agree that there are some ideas to which young children should not be exposed. However, too many parents continue to shield their children from ideas throughout adolescence and even into adulthood. They teach their children what to think rather than encourage them to learn how to think for themselves. They teach them what they believe is right and wrong rather than how to determine for themselves what is right and what is wrong. This leads to generations of people who don't know the difference between *obedience* and *morality*. Perhaps this is their goal.

Uncomfortable Idea: Parents are more interested in creating obedient children than moral children.

The online group South African Feminists state that "not all ideas are worth debating," [13] with which I would agree. I'm not spending my time debating a flat earth. Many years back, out of curiosity, I did look at the arguments and found them easily debunked. Unless a flat-earther has some new and compelling evidence that disproves hundreds of years of mathematical proofs along with geological, cosmological, and

---

[13] 6 Reasons Why We Need Safe Spaces — Everyday Feminism. (n.d.). Retrieved from http://everydayfeminism.com/2014/08/we-need-safe-spaces/

astrological evidence, it is not worth debating. The problem is with the reasons that are given as to **why** some ideas are not worth debating. The South African feminists group states that "One of our rules is that victim-blaming is absolutely forbidden; we assume that everyone in the group knows that victim-blaming is wrong." Fair enough. It's sick when people blame girls for getting raped because "they were asking for it" by the way they dressed. But is "blaming" the same thing as sharing some of the responsibility? If I go into a black neighborhood and start yelling "White people rule!" repeatedly at the top of my lungs, then I get the bejesus kicked out of me, do I not bear at least some of the responsibility for the fact that I have been beaten up even if I am not responsible for the crime of assault? We don't know how the members of this feminist group would answer these questions because these are questions that we are expressly forbidden to ask. It is my guess that this group is conflating blame for the crime with sharing some of the responsibility for the action. While it is an uncomfortable idea, there have been several studies that link provocative dress and alcohol use to rape[14][15][16] — at least enough that I, as a social scientist who understands the research, would be concerned enough about the safety of my teenage daughter to warn her how her behavior can increase the odds of becoming a victim of a crime. It sucks that we live in a society where this is true, but the fact that it sucks doesn't change the fact that it's true, and refusing to discuss this may make girls feel more liberated but certainly does not make them any safer.

---

[14] Synovitz, L. B., & Byrne, T. J. (1998). Antecedents of sexual victimization: factors discriminating victims from nonvictims. *Journal of American College Health: J of ACH, 46*(4), 151–158. http://doi.org/10.1080/07448489809595602

[15] Workman, J. E., & Freeburg, E. W. (n.d.). An Examination of Date Rape, Victim Dress, and Perceiver Variables Within the Context of Attribution Theory. *Sex Roles, 41*(3–4), 261–277. http://doi.org/10.1023/A:1018858313267

[16] Abbey, A., Cozzarelli, C., McLaughlin, K., & Harnish, R. J. (1987). The Effects of Clothing and Dyad Sex Composition on Perceptions of Sexual Intent: Do Women and Men Evaluate These Cues Differently1. *Journal of Applied Social Psychology, 17*(2), 108–126. http://doi.org/10.1111/j.1559-1816.1987.tb00304.x

Uncomfortable Idea: Victims of crimes often share some of the responsibility for the situation. Taking preventative measures (such as locking your car doors to prevent auto theft) can reduce the odds that you will become a victim.

Conscious, Individual Avoidance

Conscious, individual avoidance occurs when a person deliberately plans to keep oneself from exposure to uncomfortable ideas or decides not to entertain uncomfortable ideas. This kind of avoidance is commonly demonstrated through our preferences.

Since the beginning of human communication, we have had some level of discretion over the ideas to which we are exposed. Books, plays, public debates, movies, and radio shows all expressing ideas and exposing us to different points of view, and most of which we could choose to avoid and ignore. Prior to the mid-nineties, our media sources were extremely limited compared to the choices we have today thanks to the Internet. With what is practically unlimited choice, systems have been put in place that allow individuals to only get information from sources of their choice. Some of these systems include social media tools such as Facebook and Twitter, newsfeeds, podcasts, YouTube subscriptions, recommendation algorithms (e.g., Netflix's "because you watched X..."), and downloadable news apps. With this kind of control over the information to which we are exposed, we can easily avoid sources that expose us to uncomfortable ideas.

Uncomfortable Idea: It's very likely that your impression of the world is highly inaccurate due to how you are choosing to get your information.

One of the primary characteristics of a cult or cult-like behavior is isolation. *Isolation* includes cutting social ties with people who don't subscribe to the beliefs of the group. Some religious groups encourage and even require their members to disassociate themselves with all those outside the group. They often do this by demonizing or dehumanizing the outgroup. The popular Christian website "GotQuestions.org" makes this point clear: "Unbelievers have opposite worldviews and morals" and "The idea is that the pagan, wicked, unbelieving world is governed by the principles of Satan and that Christians should be separate from that wicked world, just as Christ was separate from all the methods, purposes, and plans of Satan."[17] The belief that nonbelievers are puppets of Satan aside, many studies in social behavior have demonstrated that intergroup contact is one of the best ways to dissolve false stereotypes, prejudice, and discrimination.[18] Of course, if a person truly believes through faith that the outgroup is evil incarnate, then attempting to persuade that person that they should seek to understand the outgroup is unlikely to be effective.

Uncomfortable Idea: If an authority figure is trying to keep you from interacting with others outside your group, that's a strong sign that you are in a cult.

Unconscious, Group Avoidance

Unconscious, group avoidance occurs when two or more people shelter themselves and/or others from uncomfortable ideas without realizing it. This kind of avoidance is demonstrated in the creation of social norms.

If one were asked to provide a few examples of racists and racists acts just ten years ago, the answers would be very different then the answers given today. Ten years ago, one might say that a KKK member is an example of a racist, and stating that Japanese Americans should "go back where they came from" would be considered an act of

---

[17] What does it mean to be unequally yoked? (n.d.). Retrieved from http://www.gotquestions.org/unequally-yoked.html

[18] Pettigrew, T. F., & Tropp, L. R. (2006). A meta-analytic test of intergroup contact theory. *Journal of Personality and Social Psychology, 90*(5), 751–783. http://doi.org/10.1037/0022-3514.90.5.751

racism. Today, one might say that Ellen DeGeneres is an example of a racist and having yourself photoshopped getting a piggyback ride from the fastest man in the world (who happens to be black), is an act of racism. This change is due to changing social norms.

Uncomfortable Idea: Racism, sexism, and bigotry are evaluations based on social norms. America is becoming more racist, sexist, and bigoted not because of changing behavior, but because of changing social norms and perception.

Defined within social psychology, *racism* requires the belief in the **superiority** of one's own race.[19] Today, however, through years of incremental and mostly unconscious processes, the definition of racism has been redefined (in the minds of the public) to include unconscious preferences for people who look like us[20] (academically known as *implicit bias*), criticism of any person of color despite the criticism being unrelated to their race, agreeing with racists on non-racist issues, eating a taco on Cinco De Mayo[21], and dressing like a person from another culture on Halloween. All these "displays of racism" may just be prejudicial, discriminatory, culturally insensitive, bad political decisions, or none of the above. Regardless, the moment we attach the label "racist" to them, we stigmatize them to the extent where even discussing the innocence of the ideas becomes an act worthy of being "racist." Social norms say that we can entertain ideas that might be prejudicial, discriminatory, and even culturally insensitive, but not "racist" ideas. Those are ideas we want to avoid appearing to support or even defend in any way.

---

[19] Nelson, T. D. (Ed.). (2009). *Handbook of Prejudice, Stereotyping, and Discrimination (1 edition)*. New York: Psychology Press.

[20] Holloway, K. (n.d.). 10 ways white people are more racist than they realize. Retrieved September 20, 2016, from http://www.salon.com/2015/03/04/10_ways_white_people_are_more_racist_than_they_realize_partner/

[21] Reporter, L. O., Reporter, T. H. P. D. M., & Post, H. (500, 17:13). Here Are 13 Examples Of Donald Trump Being Racist. Retrieved from http://www.huffingtonpost.com/entry/donald-trump-racist-examples_us_56d47177e4b03260bf777e83

Uncomfortable Idea: Your use of the term "racism" is almost certainly inaccurate. Racism requires the belief in the superiority of one's own race.

When we *romanticize*, we make something seem better or more appealing that it actually is. Think of the dozens of movies and television shows you have seen where the protagonist is faced with a decision: to keep his high-paying job where he will be away from his family for five days each week, or quit and spend lots of time with his family. We all know the ending. He quits his job, and we see the whole family spending some quality time together as happy as can be. Did you ever ask, "then what happens?" The kids eventually have to go to school, the wife has to take care of the kids, and now the father is unemployed at home all day binge watching '80s sitcoms on Netflix. The father becomes a miserable bastard with an alcohol problem; his wife leaves him, and his kids never want to see him again. The end. Admittedly, my ending is unlikely, but not any more unlikely as the romanticized endings we see on film. The reality is most often somewhere in the middle. Through expressions of culture such as movies, television, books, stories, poems, religious parables, anecdotes, and aphorisms, we unconsciously and automatically accept romanticized versions of ideas as truth. Following are some other examples of romanticized ideas. As you read this list, imagine a scene playing out describing the idea. Does it make you feel all warm and fuzzy inside? Realize that how the idea makes you feel is more of a function of the value a culture puts on the idea than any inherent goodness in the idea.

- Couples should stay together for life
- Love is always a beautiful thing
- Faith is good to have
- There is a perfect someone for everyone
- Anyone can succeed in life if they just try hard enough
- The Constitution is a blueprint for a perfect nation
- God is perfectly good as is all the advice in the Bible

- People deserve what they get in life
- Giving money to beggars is the kind and right thing to do

Unconscious, Individual Avoidance

Unconscious, individual avoidance occurs when a person shelters him or herself from uncomfortable ideas without even realizing it. Some examples of this kind of avoidance include common cultural presuppositions and committing logical fallacies.

In Islamic cultures, it is presupposed that not only a god exists, but that just one god exists, and the god's name is ALLAH, and the Koran is the word of this god. To *presuppose* something means to accept some idea as a fact without the need for critical examination of that idea. Some people call these presuppositions "self-evident truths," which ironically, are only evident to those who call them "self-evident." Often, presupposed ideas that are critically examined and found to be unworthy of acceptance could completely unravel an individual's worldview, a society, a country, and even humanity. For this reason, cultural norms protect certain ideas by demonizing outsiders who don't accept the idea, deem it "offensive" or "rude" to question the idea, and even make the disagreement or questioning of ideas illegal and in some cases punishable by death. In America, it is presupposed that "In God we trust," pledging allegiance to our country is the right thing to do, and those who kill people when our government tells them to are "heroes." The chances are, you have never given any of those ideas much thought, not because you chose not to, but because in America, these things are just a given. No critical thought required (or welcomed).

Uncomfortable Idea: Presuppositions and "self-evident truths" are ways to avoid rational justification. We need to realize that what is self-evident to us may not be self-evident to others.

Another way we unconsciously avoid exposure to ideas is by associating ideas with stigmatized people or groups, and dismissing the idea based on that association—a version of the ad hominem fallacy also known as *guilt by association*. Let's use Hitler, the classic American supervillain. How many men have you seen sportin a Hitler-style mustache? The odds are, none, outside of watching old Charlie Chaplin films. Fashion trend aside, men don't have mustaches like Hitler's because it would associate them with Hitler. For the same general reason, parents with last names of infamous people don't give their children certain first names. You don't see many young people named "Jeffrey Dahmer" or "Ted Bundy." Besides facial hair and names, we also distance ourselves from the notorious by ideas. This is seen all the time in politics when one side of the political spectrum embraces an idea, and the other side has an immediate, and unconscious aversion to the idea due to the many cognitive biases found within group psychology[22]. In short, if the idea comes from our enemy, we don't want to hear it. Unfortunately, this initial aversion combined with other cognitive biases such as *rationalization*, the *confirmation bias*, and the *backfire effect*, make the idea more uncomfortable and increasingly difficult to entertain. We'll explore these ideas and many more in the following section.

---

[22] Foss, N. J., & Michailova, S. (2009). *Knowledge Governance: Processes and Perspectives.* OUP Oxford.

Part II: Uncomfortable Ideas and the Reasons Why We Avoid Them

"It is the mark of an educated mind to be able to entertain a thought without accepting it." —**Aristotle**

We can sum up all the reasons why we avoid uncomfortable ideas with the phenomenon known as *motivated reasoning*. This describes how emotionally-charged ideas undergo a qualitatively distinct reasoning process that favors feelings over facts, which results in inaccurate conclusions and poor decisions.[23][24] Motivated reasoning has both conscious and unconscious components where both are often at play keeping us from evaluating an idea fairly if accepting that idea would be problematic for us. In this section, we'll look at some of the most compelling reasons for avoiding uncomfortable ideas and provide several examples when possible for each reason.

Unconscious Avoidance

Most of the unconscious reasons fall under the general category of *self-preservation* and are a result of the *self-serving bias*, which is any cognitive or perceptual process that is distorted by the need to maintain and enhance self-esteem, or the tendency to perceive oneself in an overly favorable manner.[25] People who think that they are horrible people are far more likely to become depressed and therefore,

---

[23] Westen, D., Blagov, P. S., Harenski, K., Kilts, C., & Hamann, S. (2006). Neural Bases of Motivated Reasoning: An fMRI Study of Emotional Constraints on Partisan Political Judgment in the 2004 U.S. Presidential Election. *Journal of Cognitive Neuroscience, 18*(11), 1947–1958. http://doi.org/10.1162/jocn.2006.18.11.1947

[24] Redlawsk, D. P., Civettini, A. J. W., & Emmerson, K. M. (2010). The Affective Tipping Point: Do Motivated Reasoners Ever "Get It"? *Political Psychology, 31*(4), 563–593.

[25] Myers, D. (2014). *Exploring Social Psychology* (7 edition). New York, NY: McGraw-Hill Education.

more likely to commit suicide than those with a positive self-image.[26] Through natural selection, the tendency to have a positive self-image has become the norm and in order make this happen, our minds have ways of deceiving us, or protecting us from reality.

The idea of the "one special someone for everyone" is compelling because going through life with self-doubt wondering if you made a bad choice because it is a statistical impossibility that you hooked up with the best person for you, is a sad way to live your life. It is easier for most people to engage in a little self-deception and believe that they found the one perfect person (out of over 3.5 billion, or 7 billion if you don't care about gender) than understand that a relationship is a collaborative effort where the couple has the power to create something close to perfection. Of course, this is a lot more work than magic. All humans are *cognitive misers* where we're constantly engaged in ways to conserve mental resources. If it is less effort to believe in the magic of fate or a divine hand, then it's more challenging to entertain the alternative ideas that require more effort. Along with alternative ideas, might come a negative self-image of a person who is stuck with the wrong person for life. And that would be bad.

Uncomfortable Idea: It is extremely unlikely that you have found the best person in the world for you.

Cognitive Dissonance

What do you think of Adolf Hitler? If I had to guess, I would say you think he was a monster. We accept that a monster can be responsible for the murder of six million Jews, but the moment we humanize this monster, we experience strong *cognitive dissonance*, or two competing thoughts or beliefs that cannot be reconciled. To us, Hitler was a monster—end of story. We don't need or want to expend cognitive energy trying to reconcile the two competing ideas that he

---

[26] OVERHOLSER, J. C., ADAMS, D. M., LEHNERT, K. L., & BRINKMAN, D. C. (1995). Self-Esteem Deficits and Suicidal Tendencies among Adolescents. *Journal of the American Academy of Child & Adolescent Psychiatry, 34*(7), 919–928. http://doi.org/10.1097/00004583-199507000-00016

was both a monster and he was also a human being with likable qualities. Accepting a simplistic half-truth is easier than accepting a complicated full-truth.

Not all monsters are results of their own actions. We create monsters by not considering positive aspects of those we are biased against in order to avoid cognitive dissonance. A person with unlikable qualities quickly becomes a monster as negative information about the person is embraced, and positive information is rejected. This process is often referred to as *demonizing*.

Uncomfortable Idea: The people you love to hate are most likely nowhere as evil as you think they are.

We all have the desire to put things in categories and reduce complexity to binary choices. This type of thinking is an example of a *heuristic*, or a mental shortcut that requires little cognitive energy and helps us to make sense of our world quickly. The problem with all heuristics is that there is a tradeoff in accuracy, and in some cases, a reduced understanding. Perhaps one of the most problematic uses of any heuristic is known as *black-or-white thinking*, which is when we force fit ideas into one of two categories. Many people base their entire worldview on this shortcut by simplifying morality into right or wrong, good or evil. And just the mere idea of "wrong" or "evil" sometimes being a matter of opinion and values is about as difficult for them to accept as idea that they are living in the Matrix.

Those with strong religious beliefs often avoid thinking about certain ideas because of cognitive dissonance. For example, how can a perfectly benevolent God allow his creations that he loves unconditionally to suffer for eternity in Hell if they don't meet the condition that they believe he exists? Some common responses to this question that illustrate the refusal to contemplate this idea is "you just need to have faith," "all your questions will be answered in Heaven," or "God works in mysterious ways." Accepting one of these answers eliminates the dissonance by eliminating the need to entertain this uncomfortable question.

Uncomfortable Idea: God really doesn't love all people, Hell doesn't really exist, and/or God doesn't exist.

Semmelweis Reflex

The *Semmelweis reflex* is a metaphor for the reflex-like tendency to reject new evidence or new knowledge because it contradicts established norms, beliefs, or paradigms. This can also be applied to uncomfortable ideas. Is it wrong to eat dogs? Reflexively, we should be horrified by the idea that people eat dogs because in most cultures, having dogs as pets is the established norm. The rejection of the idea that eating dogs is more morally wrong than eating cows or chicken is automatic and feeling-based.

Uncomfortable Idea: We can't judge people morally for eating dogs as long as we are eating cows and other animals.

The word "normal" has become a compliment and any hints at not being normal are often seen as insults. I spoke at a conference a few years back where a transgender woman gave a sincere and touching talk about her journey coming out as a transgender woman. She then opened up the floor for questions, making it very clear that people can ask anything. An elderly woman, who clearly was not aware of the social justice movement, politely asked the speaker if she feels more comfortable around people like her or around "normal women." Several members of the audience jeered and yelled: "she is normal!" Well, she is **not** normal. *Normal* is conforming to a standard; usual, typical, or expected. Transgendered men and women certainly are not conforming to any standard and only makeup about .3 percent of the population so are not at all usual, typical, nor should be expected. Out of reflex, we take offense to not being normal perhaps because the dregs of humanity are not normal, but neither were any of the greatest people in human history.

Uncomfortable Idea: Greatness, by definition, requires that we shed our desire for normalcy. If you're comfortable being normal, you're not motivated for greatness.

Overcompensation

We really don't like Nazis, and we don't want any misunderstanding in this area. If we were to learn something positive about the party, perhaps about their early efforts to discourage smoking due to the link to cancer, this might make us appear to be sympathetic to the party. We *overcompensate* for these positive feelings by publicly refusing to entertain these ideas. This is commonly done by expressing shock or horror that the idea might even be up for discussion or committing an *ad hominem* by attacking the person suggesting the idea rather than attacking the idea itself (e.g., dismissing the other person as a "bigot," "racist," "misogynist," or some other derogatory term often associated with his or her position).

Uncomfortable Idea: Name calling is often used as a way to overcompensate for the fact that the target of our verbal bashing has a valid point that we don't want to entertain.

With lawsuits, opposing lawyers don't hold reasonable positions based on the facts and fairness, they overcompensate and push for the most extreme position possible knowing that the opposing lawyer will be doing the same. This strategy works because cases are often settled through negotiation where a fair resolution can be found somewhere in the middle. In the psychology of ideology, we do the same thing for similar reasons. A *feminist* is commonly defined as one who aims to define, establish, and achieve political, economic, personal, and social rights for women that are equal to those of men.[27] This would be considered a reasonable and neutral position that we would expect every American to have in the year 2016. However, because there is a disturbing number of people (both male and female) who don't agree with this position, some feminists overcompensate by offering extreme and radical views. If a man claims that "men are better than women" the automatic and defensive response is that "women are better than men," not "neither gender is 'better' than the other." The feminists who frequently deal with anti-feminists (those holding extreme views on the issue) are more likely to become extremists for their position. We can see this today where individuals and groups advocating for man-hating, anti-marriage, all sex is rape, and female superiority, operate under the "feminist" label. But is this form of ideological overcompensation effective?

Uncomfortable Idea: There are feminists who go beyond gender equality and advocate for female superiority and other highly controversial issues, giving many people good reason not to want to associate themselves with the "feminist" label.

There is a persuasion technique closely related to overcompensation known as the *door-in-the-face technique* where large or unreasonable requests are generally rejected, but if followed by smaller more reasonable requests, the latter requests are more likely to be accepted. Imagine two groups of misogynists. The first group is asked to accept the idea that the political, economic, personal, and social rights for women should be **equal to** those of men. The second group is asked to accept the idea that political, economic, personal, and social rights for women should be **greater than** those of men, then after that idea is likely to be universally shot down, they are asked to accept the idea that the political, economic, personal, and social rights for women should be **equal to** those of men. The second group is significantly more likely to accept the proposition that the political,

---

[27] Hawkesworth, M. E. (2006). *Globalization and Feminist Activism*. Rowman & Littlefield Publishers.

economic, personal, and social rights for women should be **equal to** those of men. One point for overcompensation.

Assuming you don't accept the ideas of man-hating, anti-marriage, all sex is rape and female superiority, how would you feel about people who did hold those positions? The odds are that you would have strong negative feelings about them that you would, unfortunately, associate with the label under which they self-identify: feminism. Thanks mostly to the backfire effect, this kind of negative association does far more harm in the long run than any benefits that might be gained from effective overcompensation persuasion techniques. This is why the Westboro Baptist Church (the "God Hates Fags" people) have probably created more atheists than Richard Dawkins, Christopher Hitchens, Sam Harris, and Daniel Dennett combined.

We also see overcompensation often with atheists who reject the idea of the gods of the common religions. For example, in order to sell the idea that the God of Christianity is highly unlikely, we tend to oversell the idea that our universe is the natural result of unguided forces, that is, no god created it. But according to some—including many non-theists—it is **extremely** likely our world as we know it was created, just not by the kind of gods imagined by our ancestors thousands of years ago. The idea, made famous by philosopher Nick Bostrom in 2003, is that we are living in a simulated universe. His argument is that at least one of the following propositions are true:

1. The human species is very likely to go extinct before reaching a "posthuman" stage

2. Any posthuman civilization is extremely unlikely to run a significant number of simulations of their evolutionary history (or variations thereof)

3. We are almost certainly living in a computer simulation. It follows that the belief that there is a significant chance that we will one day become posthumans who run ancestor-simulations is false, unless we are currently living in a simulation.[28]

---

[28] Are We Living in a Computer Simulation? (n.d.). Retrieved from http://pq.oxfordjournals.org/content/53/211/243.short

So where does a god fit in all this? Let's consider one of the variations of science fiction writer Arthur C. Clark's third law: *Any sufficiently advanced extraterrestrial intelligence is indistinguishable from God*. If we create a simulated world, then we would be gods to the beings we created to inhabit the world. In other words, if we are living in a simulation, by virtually all definitions, gods exist. No matter how true this may be, it doesn't change the fact that you still need to pay your taxes.

Uncomfortable Idea: We are living in a simulated universe created by even more intelligent beings that could be considered gods.

Reaction Formation

Sigmund Freud introduced the idea of *defense mechanisms*, which are unconscious psychological mechanisms that reduce anxiety arising from unacceptable or potentially harmful stimuli, or in our case, uncomfortable ideas. One such mechanism often responsible for both avoiding and the refusal to entertain uncomfortable ideas is *reaction formation*, or the converting of unwanted or dangerous thoughts, feelings or impulses into their opposites. Although similar to overcompensation, reaction formation is a result of sympathizing with the opposition and covering up one's true feelings. We unconsciously avoid uncomfortable ideas that remind us of our lack of authenticity or the act that we are pulling a "fake-it-'til-you-make-it" scheme. People can overcompensate as a result of reaction formation, but they can also overcompensate for other reasons not having to do with actually being sympathetic to the opposition.

There is undoubtedly a number of men who identify as feminists who, primarily out of guilt of their own misogyny or sexism, are extremely vocal in the movement as a result. The refusal to entertain criticisms or respond harshly to criticisms of the movement is a result of reaction formation.

Uncomfortable Idea: A number of the most vocal proponents of minority activist groups who are not part of the minority the group represents, are motivated by guilt of their own prejudice.

One of the most common forms of reaction formation we see is conservative politicians or Christian preachers who aggressively fight against gay marriage and preach about the immorality of homosexuality, only to be caught having gay sex. This list includes televangelist Ted Haggard, North Dakota legislator Randy Boehning, former head of the Young Republicans Glenn Murphy Jr., former Republican chairman of the Cumberland County commissioners Bruce Barclay, former Idaho Republican senator Larry Craig, and Catholic priest John Geoghan who was accused of molesting over 130 young boys. This is just a partial list where the gay sex was proven either in a court of law or admitted by the perpetrator.

Uncomfortable Idea: People who are extremely vocal against homosexuality might just be that way because they are masking their own homosexual impulses.

Through reaction formation, one essentially avoids publicly entertaining an idea in which he or she sympathizes if not directly supports. This is done out of fear of exposing one's beliefs or desires about which they feel guilty or ashamed.

Intolerance of Nuance and Ambiguity

Being comfortable with or even just tolerating nuance and ambiguity is not a common characteristic found in most people. We like dichotomies: such as good/bad, right/wrong, and just/unjust. To maintain this illusion of simplicity, we create characterizations of people and ideas in our minds much like Hollywood portrays the "bad guys" in movies. Ideas that introduce realistic complexity and prove our characterizations to be inaccurate are ideas that make us uncomfortable. If we are a die-hard liberal who has characterized the

Republican Party to be one of self-serving, greedy, racists who love their guns, then we are unlikely to entertain ideas that would introduce nuance and ambiguity into our otherwise simple characterization of the party.

Consider the *Heinz dilemma*, an example frequently used in many ethics classes and classes dealing with morality.

A woman was near death from a special kind of cancer. There was one drug that the doctors thought might save her. It was a form of radium that a druggist in the same town had recently discovered. The drug was expensive to make, but the druggist was charging ten times what the drug cost him to produce. He paid $200 for the radium and charged $2,000 for a small dose of the drug. The sick woman's husband, Heinz, went to everyone he knew to borrow the money, but he could only get together about $1,000 which is half of what it cost. He told the druggist that his wife was dying and asked him to sell it cheaper or let him pay later. But the druggist said: "No, I discovered the drug and I'm going to make money from it." So Heinz got desperate and broke into the man's laboratory to steal the drug for his wife. Should Heinz have broken into the laboratory to steal the drug for his wife? Why or why not?[29]

How would you answer this question? The answer you give is not important; it's the reason you answered the way you did that is important. According to Lawrence Kohlberg's stages of moral development, the reasons fall on a continuum starting with obedience, then self-interest, conformity, law-and-order, human rights, and finally universal human rights.[30] As one moves to the right of the continuum, they accept more ambiguity and are said to have a more developed sense of morality. Many people in the United States, even adults, don't mature past the first three stages. As long as one believes without

---

[29] Kohlberg, L. (1981). *The Philosophy of Moral Development: Moral Stages and the Idea of Justice* (1st edition). San Francisco: Harper & Row.

[30] Kohlberg, L. (1976). Moral stages and moralization: The cognitive-developmental approach. *Moral Development and Behavior: Theory, Research, and Social Issues*, 31–53.

question that they have the moral high ground, any ideas that conflict with their moral judgment become far more difficult to entertain.

Is there a "right" answer to the Heinz dilemma? Or to be more specific, is there an *objectively* right answer to this dilemma that is independent of opinion? It depends on which moral theory we subscribe to. There are many moral theories such as Utilitarianism, Categorical Imperative, Aristotelian Virtue Ethics, Stoic Virtue Ethics, Consequentialism, Moral Subjectivism, Deontology, Kantian Theory, Cultural Relativism, Ethical Egoism, and others. To answer this question, we will look at just two theories: Divine Command Theory and a common alternative that uses well-being as the foundation.

Divine Command Theory. Something is good if God says it is good or if God commands it. This is an obedience-based theory. There are several presuppositions required for this theory, some of which include

- God exists
- God is good
- What God says is good, and what he commands is perfectly communicated through a book or appointed leader (e.g., the Pope)
- We have the right book (e.g., New Testament, The Holy Quran, Veda, Torah, Talmud, etc.) or we are following the right appointed leader
- The right book has been properly translated

So what does God have to say about Heinz? Unfortunately, there is no "Book of Heinz" in the Bible or any other holy book that we know of. Assuming we are Christian, we might turn to our Bible and look for "guidance" in scripture. Through a cognitive bias called *subjective validation*, we will consider a statement or another piece of information to be correct if it has any personal meaning or significance to us. So perhaps Christian #1 comes across Exodus 20:15 which reads "You shall not steal." Crystal clear as unfortunate as it may be, Heinz should let his wife die according to God. But Christian #2 recalls Proverbs 6:30 which reads "People do not despise a thief if he steals to satisfy his hunger when he is starving" and in Ezekiel 39:9-10 when

God commanded the Israelites to rob those who robbed them, so clearly God understands that there are extenuating circumstances when stealing is permitted, and what could be more godly than trying to save a life of another person? Different religions come to wildly different moral conclusions, as do different sects of Christianity, and even different Christians of the same sect have very different moral views. This doesn't say anything about the existence of a god; it just makes it very clear that what god we choose to show our obedience to, what book we choose to accept as his perfect word or leader we choose to believe speaks for this god, and all of our biases involved in answering moral questions by claiming we understand the will of this god, leads people to have very different views on morality. So according to Divine Command Theory, is there an objectively right answer to the Heinz dilemma? The answer is yes in theory (i.e., God knows the right answer), but no in practice (i.e., people come to different conclusions believing that they have the right answer from God).

What about more serious issues such as abortion? Surely God has an answer for that which most Christians can agree? Apparently not. Different religious groups support their widely varying views using the same scripture or claims of "divine revelation"—from the Catholic Church's strict opposition to abortion in all circumstances to the pro-choice stances of the Episcopal, Presbyterian, and United Methodist churches. Other major Christian and other groups have come to conclusions all over the spectrum.[31]

Uncomfortable Idea: Our morality isn't a reflection of God's; our idea of God is a reflection of our own morality.

Another aspect of Divine Command Theory is knowing what is right or wrong through direct communication with a god, either through answered prayer, the god speaking directly to the person, or

---

[31] Street, 1615 L., NW, Washington, S. 800, & Inquiries, D. 20036 202 419 4300 | M. 202 419 4349 | F. 202 419 4372 | M. (2013, January 16). Religious Groups' Official Positions on Abortion. Retrieved from http://www.pewforum.org/2013/01/16/religious-groups-official-positions-on-abortion/

through "written moral code in one's heart." Assuming that there is a god who does communicate morality this way, we still have millions of people who regularly claim such knowledge and who come to different moral conclusions. An amusing example of this schizophrenic god can be found in the 2016 Presidential election where "God" apparently has backed several Republican candidates to run for the office, according to the candidates themselves who claim to have some kind of communication with God on this issue. Ironically, one of the few Republican candidates that did not get a message from God to run, Donald Trump, won the Republican nomination.

Just because everyone thinks they know the will of God doesn't mean that nobody does; it just means that we have no way of knowing who is right.

Uncomfortable Idea: Divine Command Theory is functionally the same as everyone having their own moral standard with groups of general agreement, and everyone insisting they are right.

Secular Moral Theories. Other moral theories cannot escape many of the same problems as Divine Command Theory that has lead to different people coming to different moral conclusions. Assuming we all agree that "well-being" is the foundation or even "objective standard" of morality, we also need to agree on whose well-being we are referring. Our own? Our family's? Our country's? Humanity's? Other life forms'? Nature's? How do the groups compare in order of moral importance and how do we determine the order? Let's make another assumption that well-being applies to humanity (screw the animals[32], unless hurting animals subtracts from our well-being). But how do we value the well-being of each human? For example, are all humans the same moral value or are women and children valued more than men? Perhaps moral value is based on one's value to humanity, so a cancer researcher might be worth more than a janitor. Is a terminally-ill old man worth less than a healthy child? We can naively say that

---

[32] This is an expression. I am not advocating for beastiality.

"all life is valuable" but when put into practice, we need to know these answers to calculate well-being.

We also need to realize while well-being of humanity can be a solid foundation for morality, we don't know what well-being means to every human. Well-being itself is a subjective idea, meaning that people can have different concepts of what it means to have well-being. Well-being has been defined scientifically comprising five dimensions: Positive Emotion, Engagement, Positive Relationships, Meaning, and Accomplishment/Achievement.[33] Many different biological and environmental factors account for which of these dimensions each one of us values more, so while achievement may be most important to my well-being, positive emotions may be most important to someone else's. While we can make reasonable assumptions based on extremes, such as claiming that rape is morally wrong because it clearly lowers the overall well-being of those living in a community where rape exists, more nuanced moral choices are difficult and often impossible to know what will lead to greater well-being.

So according to well-being-based moral theories, is there an objectively right answer to the Heinz dilemma? The answer is yes in theory (i.e., there is some choice that would result in maximum well-being for all of humanity), but no in practice (i.e., people come to different conclusions based on what they think would be best for humanity).

Uncomfortable Idea: Secular moral theories have several of the same problems as Divine Command Theory, and like Divine Command Theory, cannot provide a reasonable way to **objectively** claim that one knows that something is morally right or wrong.

Given that there are differences in moral reasoning, is it acceptable to require others to follow laws based on moral reasoning and if so,

---

[33] The PERMA Model: Bringing Well-Being and Happiness to Your Life. (n.d.). Retrieved from http://www.mindtools.com/pages/article/perma.htm

when? Consider someone contemplating the Heinz dilemma and reasoning at most the advanced stage: universal human rights. The person might conclude that Heinz **should** steal the medicine because saving a human life is a more fundamental value than the property rights of another person. Or, Heinz **should not** steal the medicine, because saying this act is acceptable sets a bad precedent that would end up doing harm to humanity, which would outweigh the life of one person. When people share the same advanced moral reasoning, different conclusions can be reached based on values and perspectives. Morality is not binary, but moral issues do exist on a continuum and some issues are far more clear than others, which can reasonably justify human rights declarations such as "No one shall be held in slavery or servitude; slavery and the slave trade shall be prohibited in all their forms."

If you think you have morality all figured out or if you believe with certainty that the choices you make are the morally right ones, you are likely very wrong. All theories of morality are based on assumptions, and those assumptions are not universally shared by everyone. It would be nice if we all had an easy way to come to the same moral conclusions, but we don't. The best we can do is understand different moral perspectives and present the best possible arguments for the moral perspectives that we think are correct while being open to compromise.

Uncomfortable Idea: Morality is functionally democratic. Things are "wrong" because we generally agree they are wrong.

Feeling Over Fact

Some ideas, especially those based strongly on reason and logic, can be seen as "cold," "calculating," and "lacking humanity." When exposed to ideas like these, we become defensive and favor feeling over fact, falling victim to the *appeal to emotion fallacy*.

Is it a good idea to give money to panhandlers? Feeling tells us yes. It makes us feel good about ourselves, and perhaps even alleviates some of the guilt we have for spending $400 on that pair of alligator-skin shoes while a homeless child is digging through a dumpster looking for dinner. Although feelings tell us giving money to panhandlers is the right thing to do, the facts give us many reasons why our money is better spent by a responsible charity that serves the homeless population.[34] We need to realize that while focusing on our emotions when making decisions such as which puppy to pick from a litter is fine but focusing on reason is strongly preferred in almost all other cases, especially when we are making decisions on behalf of others. We should not avoid entertaining ideas just because they appear "cold," as these are the ideas that generally offer the greatest overall benefit to humanity.

Uncomfortable Idea: Whenever you give money to a panhandler, you are most likely doing it to relieve your own guilt or doing it because of social pressure.

Words make all the difference. Words that are emotionally charged can cause us to ignore facts, put less importance on facts, or interpret facts differently. What are your thoughts on people getting surgery to change their sex? How do you want people to feel about it? If you want people to be for it, instead of calling it "sex reassignment surgery" you call it "gender confirmation surgery." If you are not against abortion, you are "pro-choice" if you are against abortion you are "pro-life." If you're Christian, you frame becoming an atheist as a loss by referring to the transition as "losing faith." If you are an atheist, you frame the process as a gain by referring to it as "finding reason."

---

[34] Should I Give Money to Panhandlers? | The Homeless Hub. (n.d.). Retrieved September 26, 2016, from http://homelesshub.ca/resource/should-i-give-money-panhandlers

Uncomfortable Idea: We are easily manipulated by emotionally-charged terms that have a significant effect on how we see the world.

Uncomfortable and Unfalsifiable

There are many ideas that are outside the realm of science, meaning that they cannot be tested or proven to be false. These ideas are known as *unfalsifiable* ideas, some of which make us feel warm and fuzzy on the inside, and some of which are not so pleasant or are outright uncomfortable. When an uncomfortable idea cannot possibly be proven (or has not yet been proven), it gives us a reasonable excuse to ignore the idea in favor of the idea that makes us feel good.

We like to think that we are special—that the whole universe was created with us in mind. We once thought that earth was the center of the universe and that the sun revolved around the earth. In biblical times, the authors wrote as if their part of the world was the entire earth, creating stories including a "worldwide" flood, the sun freezing in the sky, and other geological events that other civilizations would have certainly noticed and recorded. To many people, the idea that the universe can contain billions upon billions of worlds comprising intelligent beings is one that adds a little humility to their reality. Because we can't currently prove that the universe is filled with life, it's more comfortable just to keep on believing that we are the purpose for the universe, not a byproduct of it. The hell with probabilities.

Uncomfortable Idea: We are most likely a byproduct of the universe and not the reason for it.

Considering how many worlds full of intelligent beings there might be in the universe, what if the multiverse theory is correct and there are 10^{500} or even unlimited universes? If you think humans are insignificant now... The multiverse theory is currently unfalsifiable, meaning specifically that we know of no way to demonstrate that this theory could be wrong, making scientific support for the theory tenuous at best. So why bother accepting such a theory that would make us feel even less special then we already are? We don't bother. We retain the comfortable idea that this is the only universe that exists and ever did exist, and we are special, dammit. Ironically, the idea that this is the only universe that exists and ever did exist is also unfalsifiable.

Uncomfortable Idea: Not only are we not alone in the universe, but our universe is not alone in the cosmos. This makes the notion that we are the reason for the universe extremely unlikely.

Protecting Sacred Beliefs

In this context, we can define *sacred beliefs* as beliefs that are off limits to criticism, doubt, or critical thought. Religious beliefs often fit into this category, but they don't have to. For example, one Christian might hold sacred the belief that divorce is morally wrong because of their reverence for the Bible. However, another Christian might be against divorce, but understand the verses on divorce to be rules for a specific audience, in a certain historical context, and willing to consider the arguments for divorce being the morally right thing to do in some situations.

Non-religious beliefs can also be held sacred, or beliefs can be held sacred for non-religious reasons. Beliefs about marriage and "appropriate" relationships are often held as sacred, even by those who don't identify with any religion, by appealing to tradition for justification. Those who argue for "traditional" marriage often claim that it has a biblical basis, but one would have to do some serious cherry-picking to make that argument. Specifically, they would need to ignore the parts of the Bible where

- Biblical marriage requires rape victims to marry their rapist (Deuteronomy 22:28-29)
- Biblical marriage requires a man to marry his brother's widow regardless of the living brother's marital status (Deuteronomy 25:5-10; Genesis 38; Ruth 2-4)

- Biblical marriage is not as good as celibacy (1 Corinthians 7:8; 28; Matthew 19:12)
- Biblical marriage involves only marrying non-foreigners. Those who marry foreigners are unfaithful to God (Ezra 10:2-11)
- Biblical marriage involves a man arranging to buy a girl from her father for an agreed upon purchase price (Genesis 29:18)
- Biblical marriage involves a wife "giving" her servant to her husband as a "wife" for sex and procreation, regardless of her maid servant's wishes (Genesis 16:2-3, Genesis 30:3, Genesis 30:9)
- Biblical marriage involves raiding villages and capturing virgins as "wives" (Judges 21:20-24; Numbers 31)
- Biblical marriage involves one man taking multiple, even hundreds, of wives and concubines (see: David, Solomon, Jacob, Abraham, and others)
- Biblical marriage involves neither partner being able to refrain from sex without the consent of the other (1 Corinthians 7:4-5)

Uncomfortable Idea: "Biblical marriage" has been described and condoned by God in many ways beyond one man and one woman. Suggesting otherwise is cherry-picking.

Proponents of "traditional marriage" also have to ignore the fact that there are countless traditions when it comes to marriage. Even if we were to focus on good 'ol American traditions, they included polygamy and bans on interracial marriage. The point is, once we see beliefs or ideas as sacred, they are impervious to critical thought. We need first to deal with the uncomfortable idea that we shouldn't be holding the idea sacred, then entertain the idea itself.

Uncomfortable Idea: "Traditional marriage" in the United States included polygamy and did not allow for interracial marriage.

If you're pro gay marriage, it's likely that you see the idea of marriage as one that evolves and don't see marriage as a sacred idea. So if we can agree as a country that gay marriage is acceptable, why not polygamy (having multiple wives) or polyandry (having multiple husbands)? How about open marriages or polyamory where either partner can have multiple partners? You might argue that these other type of marriages are immoral because they can be abusive, unfair to one or more of the partners, etc. But can't a "traditional" marriage also be abusive, unfair to one of the partners, etc.? Expanding the concept of marriage is uncomfortable to just about everyone because we hold this idea as sacred to some extent. Entertaining these ideas lead to more uncomfortable ideas we are forced to face, which makes us defensive when considering how marriage should be defined.

A common presumption is that we can only love one person romantically at a time. Is this true? You might approach that question by drawing on your own experience, and since most of us have not been involved in multiple, simultaneous relationships, we might conclude that one cannot romantically love more than one person at a time. However, we know that anecdotes can hardly be considered evidence. While the human species tends to be monogamous, the term *monogamy,* as used by biologists, doesn't exclude multiple sex partners. For example, from one study, "The term 'monogamy' does not imply lifelong exclusive mating with a single individual. In fact, many birds form pair bonds over a season, raise their offspring together, and then select another partner the following season. For biologists, monogamy implies selective (not exclusive) mating, a shared nesting area, and biparental care."[35] We also know from research that people can, and often do, romantically love more than

---

[35] Young, L. J. (2003). The Neural Basis of Pair Bonding in a Monogamous Species: A Model for Understanding the Biological Basis of Human Behavior. National Academies Press (US). Retrieved from https://www.ncbi.nlm.nih.gov/books/NBK97287/

one person at a time.[36] More research is being done in this relatively unexplored area as social acceptance for polyamorous relationships increases, and the results might be very uncomfortable to those with moral objections to such relationships. Overcoming the often sacred idea that a person can only romantically love one person at a time is a precursor to entertaining the idea that successful marriages don't have to be limited to just two people.

Uncomfortable Idea: People can romantically love more than one person at a time.

Sacredness also applies to how people feel about their country. The more hyper-patriotic we become, the more likely we are to hold certain beliefs about our country sacred. Hyper-patriotism is more accurately described as *nationalism,* which is often marked by a feeling of superiority over other countries, and by extension, the citizens of those countries. If patriotism is a form of ingroup bias, then nationalism is the nationalistic version of racism. The encyclopedia Brittanica defines nationalism as an "ideology based on the premise that the individual's loyalty and devotion to the nation-state surpass other individual or group interests."[37] As history has demonstrated, this may not be the most moral view to hold, but spun correctly, people can do horrible things while believing they are doing great things, and even "heroic" things. Many people view nationalism as a sacred idea, where others realize that countries are as fallible and imperfect as those who run them. As long as an idea is a sacred one; we are blinded by biases and incapable of making the best choices.

---

[36] Loving Two People at the Same Time. (n.d.). Retrieved from http://www.psychologytoday.com/blog/in-the-name-love/200803/loving-two-people-the-same-time

[37] nationalism | politics | Britannica.com. (n.d.). Retrieved October 1, 2016, from https://www.britannica.com/topic/nationalism

Uncomfortable Idea: Thinking your country is superior to all others is like thinking your race is superior to all others.

Conscious Avoidance

Most of the conscious reasons we avoid uncomfortable ideas fall under the general category of *fear of the consequences*. We fear that accepting the idea, entertaining the idea, and perhaps even being exposed to the idea could negatively impact our life as we know it. This in itself is a moral dilemma—do we protect people from uncomfortable ideas for their own good and the good of society, or do we discuss all ideas no matter what the consequences are? Is it better to delude ourselves when the consequences of accepting an uncomfortable truth would negatively impact our well-being?

Couples can grow apart and move on to enter relationships with different partners who are more compatible—where they are truly happy. They can still look back at their past relationship and have no regrets. They might have grown as individuals and have formed many wonderful memories that they will forever cherish. If our goal as a society is to lower the divorce rate, because we have determined divorce to be "wrong," "bad," or even "sinful," then we, as a society, don't want people accepting the idea that marriages that end in divorce can still be successful, which might remove the stigma of divorce, reduce at least some of the motivation for staying married, and provide divorcees with an alternative perspective on how to view their situation. This sounds great, but not if divorce is synonymous with failure.

Uncomfortable Idea: A marriage or long-term relationship can end in divorce/break up and still be successful.

Fear of the Slippery Slope

The *slippery slope* is both a logical fallacy and a legitimate concern. It becomes more of a fallacy the less probable the conclusion becomes. For example, "if gays are allowed to marry, then what's next... adults being allowed to marry children?" is a common, yet highly fallacious objection to gay marriage because children marrying adults is more different than gay marriage than it is similar. The slippery slope is a legitimate concern when accepting one idea brings one reasonably closer to having to accept one or more increasingly uncomfortable ideas. For example, for many Christians who take the Bible literally, accepting the idea that it is moral for two consenting adults of the same sex to love each other romantically would reasonably call into question the authority of the Bible that appears to prohibit such behavior, which could lead to the rejection of Christianity, which can lead to no longer believing in any gods, which could then unravel their entire worldview. For this reason, uncomfortable ideas that can take one down a slippery slope into even more uncomfortable ideas are often tenaciously avoided.

You have probably seen the bumper sticker that spells out the word "Coexist" using religious symbols. The idea that is being promoted is religious tolerance (also sometimes referred to as "religious freedom"). It's a wonderful idea... as long as you don't think too much about it. Once you do, you realize that there are many aspects of religion that should **not** be tolerated. Should we tolerate female genital mutilation in order to control a woman's sexual drive? How about the killing of people for criticizing their religion or trying to leave it? Or the treating women like second-class citizens? Toleration is neither a good or bad thing; it is a neutral term that only becomes a good or bad thing when we combine it with something to be tolerated. Once we agree that not every religious practice should be tolerated, we enter the slippery slope that can get us into the dangerous territory of becoming intolerant of harmless religious freedoms.

Uncomfortable Idea: Religious toleration is not always a good idea, especially when the religious beliefs result in actions that are harmful.

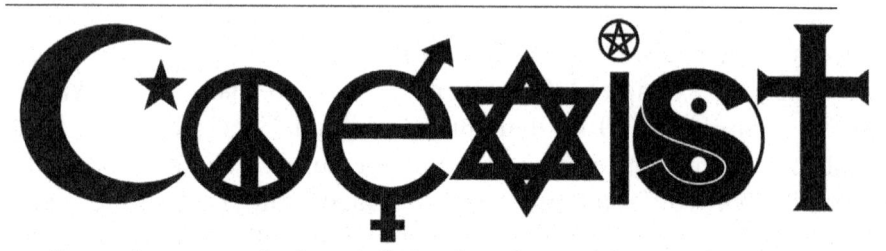

Some who promote the "coexist" idea do understand that in order to coexist, compromises must be made when pleasing one's god hurts humanity.

Fear For Society

Take the uncomfortable idea that freewill is just an illusion, and every choice and decision we make is a result of deterministic or random environmental and biological factors. What would happen if everyone believed this? In one study, those who hold the belief that there is no freewill were more aggressive and less likely to help others.[38] In other studies, those who feel like they have a sense of control over life situations are generally healthier and have a greater sense of well being.[39] However, research has also demonstrated that high belief in freewill has been linked to a punitive attitude toward

---

[38] Baumeister, R. F., Masicampo, E. J., & DeWall, C. N. (2009). Prosocial Benefits of Feeling Free: Disbelief in Free Will Increases Aggression and Reduces Helpfulness. *Personality and Social Psychology Bulletin, 35*(2), 260–268. http://doi.org/10.1177/0146167208327217

[39] Lachman, M. E., & Weaver, S. L. (1998). The sense of control as a moderator of social class differences in health and well-being. *Journal of Personality and Social Psychology, 74*(3), 763–773. http://doi.org/10.1037/0022-3514.74.3.763

wrongdoers and lower forgiveness toward them.[40] Accepting ideas have consequences, as does rejecting them.

Uncomfortable Idea: Freewill is just an illusion, and every choice and decision we make is a result of deterministic or random environmental and biological factors.

Even if there were some magic known as freewill, this wouldn't change the fact that our decisions are at least strongly influenced by outside forces. One of the best examples of this can be found with organ donation. In countries where organ donation forms have *presumed consent* (or an opt-out checkbox), organ donation increases around 30%.[41] While some of this difference can be due to the fact that people did not read the form, research has shown that perception of the norm has a strong influence on people's decisions—even ones as significant as organ donation. This is known as the *bandwagon effect*. If you think most people donate their organs (as inferred by the default option), then you are more likely to donate your organs. Social psychologists know this trick and many more that "nudge" behavior. Marketers, politicians, salespeople, and others use these strategies to their advantage, and sometimes to your detriment.

---

[40] Baumeister, R. F., & Brewer, L. E. (2012). Believing versus Disbelieving in Free Will: Correlates and Consequences: Free Will Beliefs. *Social and Personality Psychology Compass*, *6*(10), 736–745. http://doi.org/10.1111/j.1751-9004.2012.00458.x

[41] Rithalia, A., McDaid, C., Suekarran, S., Myers, L., & Sowden, A. (2009). Impact of presumed consent for organ donation on donation rates: a systematic review. *BMJ*, *338*(jan14 2), a3162–a3162. https://doi.org/10.1136/bmj.a3162

Uncomfortable Idea: As much as you like to think you are in complete control over your decisions, behavior, and actions, you're not. We are social beings, and we all influence each other—none of us live in a social vacuum.

Consequences can be measured scientifically, either experimentally or statistically. They can also be imagined. The religious often avoid critical discussions about their beliefs because they fear a world where people's behavior isn't motivated by eternal punishment or reward. While there have been studies demonstrating that people behave more ethically and prosocially when they are reminded by religious symbols,[42] the same ethical behavior is seen when people are shown pictures of eyes looking at them,[43] or when they see themselves in a mirror.[44] The strongest evidence against the imagined fear that without belief in God, chaos will ensue, is the fact that it doesn't. On virtually all related measures, secular societies are better off than religious ones notably in terms of less crime, less poverty, and overall well-being (happiness).[45] If the evidence is strongly against your imagined consequences, then your fear is unreasonable.

---

[42] Shariff, A. F., & Norenzayan, A. (2007). God Is Watching You Priming God Concepts Increases Prosocial Behavior in an Anonymous Economic Game. *Psychological Science, 18*(9), 803–809. http://doi.org/10.1111/j.1467-9280.2007.01983.x

[43] Bateson, M., Nettle, D., & Roberts, G. (2006). Cues of being watched enhance cooperation in a real-world setting. *Biology Letters, 2*(3), 412–414. http://doi.org/10.1098/rsbl.2006.0509

[44] Beaman, A. L., Klentz, B., Diener, E., & Svanum, S. (1979). Self-awareness and transgression in children: Two field studies. *Journal of Personality and Social Psychology, 37*(10), 1835–1846. http://doi.org/10.1037/0022-3514.37.10.1835

[45] Secular Societies Fare Better Than Religious Societies. (n.d.). Retrieved September 24, 2016, from http://www.psychologytoday.com/blog/the-secular-life/201410/secular-societies-fare-better-religious-societies

Uncomfortable Idea: Societies will not erupt in chaos if people don't believe in eternal punishment or reward.

Let's take a cynical and perhaps conspiratorial look at the United States military. National defense is a big business, with the US spending more on defense than the next top-spending seven countries combined.[46] That dollar amount needs to be justified to the American taxpayers. Joining the military is no dream job—the pay is not great; it's dangerous and risky; many of those who do return from war will struggle with post-traumatic stress disorder, anxiety, and depression; and veterans will be twice as likely to commit suicide than the average civilian.[47] If the government hired me as a marketing consultant to address these issues, and I had no morals, I would advise them to do the following:

- Create an advertising campaign glorifying battle and highlighting the really cool stuff (like jumping out of planes).

- Demonize and dehumanize the enemy. Tell the American public that our enemy "hates us for our freedom." Make all conflicts as black and white as possible where our enemies deserve 100% of the blame. It takes a lot of motivation for the average person to take another human life.

- Exaggerate and sell fear. Tell the American public that our way of life is being threatened.

- Stop using terms like "soldiers" that implies killing, and start using a more image-friendly term like "troops," which elicit images of men and women in uniform giving little kids high-fives.

---

[46] U.S. defense spending compared to other countries. (n.d.). Retrieved from http://www.pgpf.org/chart-archive/0053_defense-comparison

[47] Relations, M. (2016, May 12). Facts about Veteran suicide. Retrieved from http://www.blogs.va.gov/VAntage/27677/facts-veteran-suicide/

- Associate supporting the military with patriotism and criticizing it with being unpatriotic.
- Since people are more motivated by social currency such as gratitude and respect than they are motivated by money, brand all military personnel as "heroes" and dress them up in snazzy uniforms so they can easily be identified.
- Work closely with religious institutions and integrate religion into the military so Jesus' command of "turn the other cheek" can be interpreted to mean "kill the people that you're ordered to kill."
- Make killing easier by distancing the one pulling the trigger from the humans they kill.

My hypothetical advice probably resembles actual advice once given to our government. But this is the kind of idea we don't want people finding out that it's true because of the consequences it will have on society. One can argue that the American people are victims of *prosocial manipulation*, and the manipulation is for the good of the country (we wouldn't last long with a "turn the other cheek" approach). After all, without the bad guys, we can't be the good guys, without the guts, there'd be no glory, and without the approval from the churches, all this killing would be Satan's work, not God's.

Uncomfortable Idea: The American people are being manipulated to be more pro-military, to hate the enemy, and to want to kill. But this manipulation is necessary for national security.

We Don't Want To Be Seen As "Unpatriotic"

Patriotism is defined as "an emotional attachment to a nation which an individual recognizes as their homeland." This is a highly subjective term, and patriotism can be expressed in many ways. People can also act in ways that are interpreted as unpatriotic when in fact their actions are a result of their emotional attachment to their nation. Being seen as "unpatriotic" is virtually always negative and actively avoided.

Despite the fact that the Founding Fathers built a mechanism into the Constitution to amend or change the Constitution, and despite the fact that "we the people" have exercised this mechanism 27 times, many people see it as unpatriotic to question the Constitution. In fact, "unconstitutional" is synonymous with "wrong." We forget that the Founding Fathers wrote the Constitution in a political, technological, and social climate very different from the one we live in today. A common example presented these days is the "right to bear arms" amendment that was written in a time when "arms" were very different than the "arms" we have today. The Constitution isn't perfect, and we have a right and responsibility to point that out.

Uncomfortable Idea: The Constitution is not perfect.

Would we want the Founding Fathers running our country today? Although many hail them as gods, they were not. They were politicians with the imperfections and flaws of politicians today. Although it would be unfair to judge their actions and beliefs in today's political and social climate (this is known as the *historian's fallacy*), if we did, it would be very clear that these are not people we want running our country today—unless we're okay with our leaders owning slaves and denying voting rights to blacks and women.

Uncomfortable Idea: The Founding Fathers were not perfect, and we would not want them running our country today. Disagreeing with them doesn't make you unpatriotic nor does it make you wrong.

Most people don't realize (or don't want to know) that the "traditional" Pledge of Allegiance did not contain the words "under God." For 62 years, Americans were reciting the Pledge while keeping religion out of government. In 1954, after more than a half century of reciting the Pledge and despite the objection of Francis Bellamy's daughter (Francis was the minister who wrote the Pledge), President Eisenhower encouraged Congress to add the words "under God" to differentiate us from the Communists at the time.[48] If there is a God, then all countries are "under God" and the addition is as pointless as saying "under clouds." If the "under God" implies this God shows us special favor, this is simply an unwarranted claim made by every other country that believes that they have a god or the gods on their side. What was once a pledge that unites us, now divides us. Nearly one in ten Americans claim not to believe in any gods, and that number is significantly larger for younger generations.[49] **If we insist on associating patriotism with religion, the less religious we become, the less patriotic we become.** It is a catch-22 because religion is associated with patriotism, suggesting that it shouldn't be is seen as being "unpatriotic," and thus an uncomfortable idea we like to avoid.

Uncomfortable Idea: The phrase "under God" was added to the Pledge of Allegiance 62 years after being recited without it. It should be removed because it links patriotism to belief in a monotheistic God, which a growing number of Americans don't have.

What does it even mean to pledge our allegiance to our flag? This is essentially making a promise to our country, but a promise of what, exactly? The word "allegiance" can mean loyalty, dedication, devotion, fidelity, honor, obedience, or homage. All of these terms are comparative terms, that is, without some object of comparison or baseline behavior, they have no meaning. For example, would you promise to be loyal to your country? Does this mean that if your country asked you do something that strongly contradicted your sense of morality, you would? If your country's laws required you to sit at the back of the bus because of the color of your skin, would you pledge your obedience to your country and follow the law? Would you honor your country for the decision their leaders made, like starting the

---

[48] The Pledge of Allegiance. (n.d.). Retrieved from http://www.ushistory.org/documents/pledge.htm

[49] Lipka , Michael (2015, November 4). Americans' faith in God may be eroding. Retrieved from http://www.pewresearch.org/fact-tank/2015/11/04/americans-faith-in-god-may-be-eroding/

Iraq war? Or is your allegiance to your own sense of morality whether you think it is a reflection of your god or your sense of empathy combined with some form of consequential ethics? Even with the Marines, "God" comes before "Country."

So what's the point of the Pledge? The repetition of Pledge is an affirmation that has powerful effects on how we unconsciously feel about that which we are affirming. This can be a good thing for our country as it will cause people to be more likely to put their country before self-interests even including the religious and moral ones. This is similar to the evolutionary idea that altruism exists because it is a successful strategy for the survival of the species even though individual sacrifices are made. Not good for the individual, but good for the group.

Uncomfortable Idea: Reciting the Pledge conditions citizens to be obedient and loyal to their country. It is another example of prosocial manipulation.

American children performing the Bellamy salute while reciting the Pledge of Allegiance. The salute was replaced with the hand over the heart shortly after World

War II because the salute had too much resemblance to the Nazi salute. Photo public domain circa 1940.

Is America the greatest country in the world? First let's look at the word "great," which means "of an extent, amount, or intensity considerably above the normal or average." Considerably above the normal or average in what? In "countryness"? If the question were seriously asked, "what is the greatest country in the world," we would need to *operationalize* the term "great" by associating it with certain metrics, in other words, it would have to be measurable. So for example, we can say that for our purposes, "greatness" is defined by the average score of a collection of over 50 measurable criteria including crime rate, reported happiness of citizens, employment rate, racism, safety, human rights, etc. The *U.S. News and World Report* recently ranked the countries using this general method.[50] By their metrics, the United States came up fourth, trailing the United Kingdom, Canada, and Germany. Through what is known as the *Texas sharpshooter fallacy*, one can easily look at all the collected data on all the countries, then choose to define "great" or "best" by cherry-picking metrics that their country scores highest on, and leaving out the others. We can make any country the "greatest." There is no universally agreed upon set of metrics for greatness since greatness is largely defined by values, and values differ greatly among cultures.

But what about people who just love their country so much that they really do believe it is the greatest in the world? This is like the "world's greatest dad" claim, you either don't really believe it and just say it because you think it's patriotic to say, or you really do believe that our country is better than every other country in the world, which given the overwhelming evidence against such a claim using any reasonable number of metrics, makes you sound like someone who is deliberately ignoring the data to maintain a sense of superiority over others.

---

[50] The 60 Best Countries in the World. (n.d.). Retrieved October 3, 2016, from http://www.usnews.com/news/best-countries/overall-full-list

Uncomfortable Idea: America might be great, but it is not the greatest country in the world.

"If you don't love your country, then get out!" is not an uncommon response to any criticism made toward the country or admission that America isn't the "greatest country in the world." First of all, just because you think something isn't the greatest doesn't mean you don't want it. My car is awesome, but there are many better cars available. I am still grateful to be driving it. Second of all, unless America is perfect, we need people who can call attention to the problems that the rest of us conveniently ignore, especially if those people offer and initiate proposed solutions.

Uncomfortable Idea: America needs leaders who care enough about the country to make it better, not leaders who make inaccurate claims about its greatness to appear patriotic.

"God Bless America" doesn't make much sense. Of course, as a secular American, I would prefer that our elected leaders didn't publicly ask favors from invisible beings on behalf of all Americans. But let's work under the assumption that the Abrahamic God of theism and the Bible exists, and he does take special requests. Regardless of what being "blessed" actually means, it's clear that it's something good that God gives. Let's even ignore the theological assumption that God knows what is best for everyone and doesn't need us telling him what to bless. We are essentially asking an all-powerful God to do something for America and not for other countries. This isn't the same as screaming "All Lives Matter" in response to "Black Lives Matter," because we are creatures of limited focus, attention, and resources who need to prioritize. But the God of theism does not have these limitations (if he did, he wouldn't be God). Wouldn't the world—and America—be better off if God "blessed" every country, political faction, militia, and person in the world? Shouldn't God also bless ISIS, so they realize that throwing gays off buildings isn't cool? What about African children who are currently dying from starvation at a rate of 3.1 million per year? Asking an unlimited God to limit his blessings to our already highly-privileged country doesn't make sense. If we really did believe in an all-powerful God, who granted special favors upon request, why wouldn't we ask this God to "bless" humanity instead? Maybe "God bless America" is nothing more than a necessary catchphrase to get and maintain political support in a dominantly religious culture.

Uncomfortable Idea: "God bless America" is a stupid thing to say even for American theists.

The Desire to Hold Popular Views or the Fear of Social Response

Social interaction and relationships have been shown to be one of the greatest indicators of well-being. We maintain positive relationships by being agreeable and not argumentative or contrarian. Rather than sacrifice our relationships, we choose to avoid entertaining ideas that would put us in the position of having to choose between our relationships and our beliefs and ideals. This makes dismissing uncomfortable ideas easy.

Social media feeds have been called *echo chambers* because people have a strong preference to follow or friend others who share similar views when such views are important to them. In line with *operant conditioning* in psychology, we are conditioned to hold and share ideas when the sharing of such ideas are rewarded rather than punished. The showering of likes, supportive comments, and shares shape our behavior more than we realize. We often dismiss uncomfortable ideas publicly as a way to get more rewards and avoid punishment. This problem is magnified due to *polarization*.

Uncomfortable Idea: Our beliefs and opinions are formed largely by the response of others through punishment and rewards such as hostile responses and social bonding through agreement, respectively.

Imagine there is breaking news for the first time about a white cop who shot and killed a black suspect. Liberal media outlets might focus on why the shooting was unjustified, and conservative media outlets might focus on why the shooting was justified. Assume the facts of the case allow for a reasonable interpretation either way. As people start to comment on the story within social media, more varied opinions are offered, but those who express opinions that support one view are rewarded more than those expressing other views. This view quickly becomes the dominant view that is most likely to lead to reward in the form of likes and supportive comments. As time passes, moderate support is common and no longer rewarded as it once was, so more extreme views are needed to elicit the same level of reward. Moderate comments such as "I can see both sides of this issue" are buried by top comments that might read "Racist cops should be burned at the stake!" or "If you run from a cop you deserve to get shot. End of story!" As the average group position on the matter becomes more extreme, it becomes increasingly difficult for a member of that group to hold a contrarian view. The contrarian view might come to be held as a result of entertaining an uncomfortable idea. For liberals, the uncomfortable idea might be the idea that the cop was justified in shooting the suspect, whereas for conservatives it would be that the cop was unjustified.

Uncomfortable Idea: Your perception of racist cops and justified shootings by police is strongly influenced by your politics and how you get your "news."

In the age of equality, people like to pretend that everyone is the same in more ways than is confirmed by reality. Even suggesting that men and women are different is often enough to be a social faux pas worthy of gasping. We all can agree that biologically, there are physical differences between the sexes, some more obvious than others. But that is where most people draw the line of admission. Biology also comprises chemical hormones (e.g., estrogen and testosterone) that have drastic effects on behaviors, sexual drives, and emotions. Environmental factors aside, these biological factors are powerful enough to make a significant difference in how each of sexes might perform tasks. For example, as a group, men are better suited for jobs that have greater physical demands that require strength. Women, as group, are better suited for jobs that require empathy.[51] It is important to point out that these are group generalizations made from the average. This means that any given male could have more empathy than the average female, or any given female could be stronger than the average male. It is not at all unreasonable to conclude that, as a general rule, some jobs are better suited for men, and some are better suited for women. This wouldn't be a conclusion based on culture, the patriarchy, or sexism; it is a conclusion based on biological facts.

Uncomfortable Idea: Men and women are different biologically, which means there are differences in emotions, drives, cognition, and behavior.

You might have heard that "race doesn't exist" or that "race is a social construct," implying that the Chinese, Swedes, Indians, and Africans are all the same. Having similar human characteristics does not mean that people are all the "same." Race is a social construct in that race exists because we say it exists. It is also a categorical system and like all other categorical systems, fails to perfectly demarcate the categories. There are clearly white guys, and there are clearly black guys, but there are also many people in between. "Pure races" never existed. Biological differences account for some of the physical markers that we use to determine race such as skin color, eye shape, length of limbs, hair color and thickness, and other more subtle cues.

---

[51] Mestre, M. V., Samper, P., Frías, M. D., & Tur, A. M. (2009). Are Women More Empathetic than Men? A Longitudinal Study in Adolescence. *The Spanish Journal of Psychology, 12*(1), 76–83. http://doi.org/10.1017/S1138741600001499

These differences reflect both hereditary factors and the influence of natural and social environments.[52] Racial differences can be important because there are medical differences among races that are vital to acknowledge in treatment and disease prevention. With very few exceptions, however, race is perceived to be important when it is not at all important. This is the basis of prejudice and the reason why the "race doesn't exist" people don't like acknowledging the biological differences that really do exist.

Uncomfortable Idea: Race may be a social construct but biological differences in races do exist.

Privilege is a social theory where special rights or advantages are available only to a particular person or group of people. While the concept is an important one within social justice, "privilege" is a term flippantly used by members of every group who like to point out that they have it worse than members of other groups. The term "privilege" is often misleading because it implies that the privileged group is getting some benefit they don't deserve, rather than what is often happening: **the underprivileged group is not getting the fair treatment they do deserve**. We don't want to treat the privileged worse; we want to treat the underprivileged better. Take for example the gender pay gap in the United States. Without citing specific figures, it is very clear that some portion of the difference men and women get paid for the same work is due specifically to gender, which can be framed as "male privilege in the workplace." The problem isn't that men are getting paid too much; it's that women aren't getting paid enough. Correcting this is an example of a *positive-sum outcome*, where the total of gains and losses is greater than zero.[53]

---

[52] Biological Aspects of Race. (n.d.). Retrieved from http://physanth.org/about/position-statements/biological-aspects-race/

[53] positive-sum game | game theory | Britannica.com. (n.d.). Retrieved October 19, 2016, from https://www.britannica.com/topic/positive-sum-game

Uncomfortable Idea: Underprivileged folks, if you frame the privilege in such a way that puts blame on the privileged group or threatens to take away the alleged privileges, it is likely to be met with hostility and resistance. This is a common and expected response. Be prepared for it, and get used to it.

Unfortunately, there are privileges based on a *zero-sum outcome*, meaning that to address the social injustice, the privileged group has to give up that which they are unjustly getting. An example of this can be found in racial preference for hiring—a phenomenon that can be accurately described as "white privilege in the workplace." In a 2003 landmark study, researchers sent fictitious resumes to help-wanted ads in Boston and Chicago newspapers, manipulating the perceived races of the applicants by randomly assigning Black- or White-sounding names to identical resumes. White names received 50 percent more callbacks for interviews.[54] The number of available jobs are limited so if this injustice is corrected it would mean that whites will get fewer callbacks thus fewer jobs. In the larger scheme, correcting such social injustices is a positive-sum game because overall, societies based on fairness and justice do better. However, asking people to give up personal benefits for a greater social good is never an easy sell.

Uncomfortable Idea: Privileged folks, sometimes you will experience personal losses when injustice is addressed. Get over it. It's for the greater good.

Individual members of privileged groups are often held responsible for unfair social norms and laws that have existed sometimes for centuries. The justification for this responsibility is summarized in the quote by Eldridge Cleaver, "There is no more neutrality in the world.

---

[54] Bertrand, M., & Mullainathan, S. (2004). Are Emily and Greg More Employable Than Lakisha and Jamal? A Field Experiment on Labor Market Discrimination. *The American Economic Review*, *94*(4), 991–1013. https://doi.org/10.1257/0002828042002561

You either have to be part of the solution, or you're going to be part of the problem." If you are a white woman not fighting for the rights of black men and women, then you are part of the problem. If you are a black man not fighting for the rights of black or white women, then you are part of the problem. If you are a black woman not fighting for the rights of the physically disabled, then you are part of the problem. No matter who you are, according to the above oft-cited quote, you are part of countless problems. There are literally hundreds of groups that are somehow less privileged than you are, and if the privileged are responsible for the less privileged, then you are guilty. This is guilt that you can't possibly make go away because it is impossible to do something for everyone who has it worse than you. Those who realize this often resort to self-loathing and guilt because of their privilege. By publicly expressing guilt and shame, one can share in the suffering of the less privileged. Guilt and shame sometimes work because they are powerful motivators, but so is a well-reasoned desire for fairness and a passion for humanity.

Uncomfortable Idea: Acknowledge your privilege and show empathy to those who don't enjoy the same privilege. Instead of tearing yourself down with guilt, shame, and self-loathing, bring others up by helping them get the fair treatment they deserve.

2016 was an active year for "bathroom bills" and transgender rights. On the one side, you have transgender people who want to use the gendered public bathroom with which they identify. On the other side, you have those who want people to use the bathroom that corresponds to the sex on their birth certificate. There are so many problems with the latter position I won't even bother to address them, since the vast majority of Americans understand them. But what about people who are born male, who now identify as female, but don't want to change their appearance and in some cases even have beards? It's not difficult to understand why a woman would feel unsafe when seeing someone who looks unquestionably like a stereotypical man in the woman's bathroom. Is it okay for masculine-looking people with beards and penises to use women's bathrooms just because they identify as women? It is not an easy question to answer, but that's the point. We cannot call people who raise these questions "transphobic" and expect the issues to go away. We shouldn't avoid raising the questions because we don't want to be called a bigot. The questions need to be asked, discussed, debated, and only then can they be answered.

Uncomfortable Idea: Transgendered women with beards should stay out of women's bathrooms.

There has been an effort to redefine racism and sexism to require power; that is "prejudice plus power." This is a fringe definition at best and certainly does not accurately represent the majority of uses within the social sciences. Prejudice plus power leads to *systemic racism* and *systemic sexism,* legitimate terms that only focus on one manifestation of racism or sexism at a time within a system. Official definitions are descriptive in that they are frequently updated to describe common usage and nuance. One day, racism and sexism might be redefined in Merriam-Webster, the Oxford Dictionary, and the Cambridge Dictionary, as well as peer-reviewed social psychology and sociology textbooks. But for now, these sources simply have no mention of power being a requirement.[55][56][57] Regardless of how we play with language, words still represent ideas. Even if we make it impossible for non-whites to be racist or women to be sexist, this won't change the fact that there are non-whites and women who are prejudice, discriminatory, or antagonistic against someone of a different race or sex based on the belief that their own race or sex is superior. Nobody gets a free pass for being an ignorant ass because of the shade of their skin or their gender.

---

[55] racism Definition in the Cambridge English Dictionary. (n.d.). Retrieved October 4, 2016, from http://dictionary.cambridge.org/us/dictionary/english/racism

[56] racism - definition of racism in English | Oxford Dictionaries. (n.d.). Retrieved October 4, 2016, from https://en.oxforddictionaries.com/definition/racism

[57] Definition of RACISM. (n.d.). Retrieved October 4, 2016, from http://www.merriam-webster.com/dictionary/racism

Uncomfortable Idea: Non-whites can be racist, and women can be sexist.

On a related note, I have come across many atheists who argue that we can't use the dictionary definitions of racism or sexism and we have to use their understanding of the word instead. These are the same atheists who rage against Christians who insist that atheism means "insisting there is no god," despite the fact that virtually every dictionary and encyclopedia definition defines atheism as a "disbelief in a god or gods" or at least includes that in its primary definition. Atheists, you can't have it both ways.

Recently we have seen an explosion in *trigger warnings*, or "a statement at the start of a piece of writing, video, etc., alerting the reader or viewer to the fact that it contains potentially distressing material." Although mostly used in college courses, you can now find trigger warnings at all levels of the educational system. Those who support the use of trigger warnings do so because it allows students who might have been traumatized to prepare themselves mentally for encountering material that might "trigger" strong negative emotions related to the trauma so they can better manage their reactions. Those who oppose trigger warnings argue that trigger warnings allow students to avoid exposure to information they shouldn't be avoiding either for educational reasons or because the information will help desensitize their triggers. Regardless of the overall pros and cons of trigger warnings, we should look at the triggers themselves.

Our responses to other people's traumatic events play a large part in how traumatic they see the events. If you are a parent, you realized at some point that if you shower your child with attention for every bump and bruise, the bumps and bruises become more frequent and more "traumatic." This is standard human behavior reinforced by reward, mostly through attention and sympathy. While a sexual assault by a drunk uncle, watching your friend get shot, or a death of a loved one are not on the same level as a bump or a bruise, the psychological principles still apply. We might shower others with attention and sympathy for their traumatic events because we are extremely empathetic people and it is our natural and sincere response, or we are just responding how we think we should respond—the socially appropriate response. While the sympathetic response, dripping with emotion, may be the nice response, the kind response considers the long-term well-being of the individual, even at the expense of coming across as insensitive.

How come some people develop full-blown *post-traumatic stress disorder (PTSD)* after a negative life event when others who experience a similar life event do not? There are many factors that could account for this including genetic and environmental ones, some of which are outside of our control and others that are within our control. We can deal with our own trauma through something called appraisal theory. *Appraisal theory* is the claim that emotions are the result of a person's subjective evaluation or significance of a situation.[58] This means that we do have a level of control over how we respond to unfortunate life events. Let's say a meteor hits the earth and kills everyone we know. We will have a physiological reaction to this event that is extremely difficult, if not impossible to control. This reaction may include increased heart rate, trembling, sweating, shortness of breath, and *primary* or *basic* emotions such as fear and sadness. It is at this point our appraisals kick in. Through a cognitive process, we evaluate our situation and here is where we have more control over our emotions moving forward. Perhaps we focus on how grateful we are that we are still alive and focus on a new life-purpose of finding survivors and rebuilding civilization. Or we can focus on how horrible everything is, slip into depression, then jump off what is left of the Brooklyn bridge. This isn't always an easy choice to make, but it is a choice.

---

[58] Scherer, K. R., Schorr, A. E., & Johnstone, T. E. (2001). Appraisal processes in emotion: Theory, methods, research. Oxford University Press.

Uncomfortable Idea: Trauma can become a self-fulfilling prophecy when negative life events are treated as traumatic events, and sympathy and attention positively reinforce beliefs that one should be emotionally devastated.

Gays don't have it easy. They are discriminated against, judged by many to be immoral, targets for bullying, far more likely to experience life-threatening sexually-related diseases;[59] more likely to suffer from mental illness,[60] and more likely to commit suicide.[61] No parent in their right mind would want their child to have to deal with these issues. Being upset when a son comes out as gay is an appropriate reaction, providing the parents don't blame the son for being gay. This does not make the parents bigoted, homophobic, or anti-gay; it makes them loving parents who, like all parents, want the best for their children.

Uncomfortable Idea: Non-bigoted and non-homophobic parents don't want their sons to be gay.

The "gay agenda," as some like to call it, includes normalizing homosexuality, which many fear will "spread homosexuality" and make it more common within a population. It can't be denied that activists for gay rights do want to normalize homosexuality in the sense that being gay is no longer stigmatized... not by making everyone gay. "Normal" in this context refers to the kind of behavior

---

[59] Gay and Bisexual Men | HIV by Group | HIV/AIDS | CDC. (n.d.). Retrieved October 7, 2016, from http://www.cdc.gov/hiv/group/msm/

[60] J. C. ~ 3 min. (2016, May 17). Higher Risk of Mental Health Problems for Homosexuals. Retrieved from http://psychcentral.com/lib/higher-risk-of-mental-health-problems-for-homosexuals/

[61] Facts About Suicide. (n.d.). Retrieved from http://www.thetrevorproject.org/pages/facts-about-suicide

and not the frequency of a behavior. It should also be noted that the phrase "spread homosexuality" and other similar phrases are used to elicit fear in the general population with the implication that you just might catch gayness. People don't catch gayness, but normalizing homosexuality is likely to result in more people identifying as gay as well as more people having gay sex.

Uncomfortable Idea: Normalizing homosexuality is likely to result in more people identifying as gay as well as more people having gay sex.

This should not be a controversial idea. Assuming legalized gay marriage is a reasonable indicator of normalization, the countries that have legalized gay marriage tend to report higher percentages of citizens identifying as gay, lesbian, or bisexual. Not surprisingly, there are no self-identified gays in countries where homosexuality is punishable by death. At least none left. The reasonable inference from this data is that social acceptance allows people to be open and honest about their sexuality. But if being gay is not taboo, are some people more likely to choose to be gay, like choosing rocky road over rum raisin ice cream?

Although I have written extensively on the question if people are born gay in my book, *Reason: Book I*, for this point, it doesn't matter if people are born gay or not. What does matter is that it is understood that sexual attraction is not a binary construct, but a feeling that exists on a continuum ranging from exclusively attracted to the opposite sex to exclusively attracted to the same sex. There are many people who hover in the middle of this continuum who often identify as "bisexual." These are people who find both men and women sexually attractive and can choose to pursue relationships with either sex. Again, it should not be a controversial idea that bisexuals in societies where homosexuality is stigmatized are more likely to pursue heterosexual relationships. "Gay" can be used to represent the **feeling** of attraction toward someone of the same sex, sexual **behavior** with a partner of the same sex, or the **label** one uses to self-identify. One cannot choose a feeling since a feeling (known as *affect* in psychology) does not require cognition—it is a basic physiological response to our environment. But one can choose the kind of behavior in which they engage as well as the label they use to self-identify. So if by "gay" we are referring to behavior or the label, then yes, one can choose to be gay.

Uncomfortable Idea: Some people do "choose to be gay."

We don't know exactly why some people are sexually attracted to people of the same sex. Homosexuality does have a heritable component.[62] This means that we can attribute at least some of the same-sex attraction to genes if not all of it. Gene expression is often the result of interaction with environmental factors, so developing same sex attraction could be partly due to early life experiences. The official statement from the American Psychological Association (APA) reads:

There is no consensus among scientists about the exact reasons that an individual develops a heterosexual, bisexual, gay, or lesbian orientation. Although much research has examined the possible genetic, hormonal, developmental, social, and cultural influences on sexual orientation, no findings have emerged that permit scientists to conclude that sexual orientation is determined by any particular factor or factors. Many think that nature and nurture both play complex roles; most people experience little or no sense of choice about their sexual orientation.[63]

This possibility is not a license to make anecdotal claims such as "uncle john touched my private part once, so now I am gay." People generally make horrible armchair scientists when attempting to figure

---

[62] Pillard, R. C., & Bailey, J. M. (1998). Human sexual orientation has a heritable component. *Human Biology, 70*(2), 347–365.

[63] Answers to Your Questions For a Better Understanding of Sexual Orientation and Homosexuality. (n.d.). Retrieved October 7, 2016, from http://www.apa.org/topics/lgbt/orientation.aspx

out causes of their own behavior. But just because we can in no way make reasonable claims that bad relationships with parents, playing with dolls as a child, or being inappropriately touched causes or even contributes to same-sex attraction, scientifically speaking, we cannot rule it out.

Uncomfortable Idea: We cannot rule out the possibility that environmental factors can result in same-sex attraction.

Normalizing homosexuality will likely result in more people identifying as gay, and more people willing to engage in same-sex behavior. It is possible that it also might result in more people having same-sex attractions—we just don't know as it would be extremely difficult, if not impossible, to measure. While normalizing homosexuality would appear to put more people at risk for harm, it is important to realize that much of this harm is directly related to the stigmatization of homosexuality that normalizing would reduce or even eliminate. From a humanitarian standpoint, we also need to realize that it is very likely a significant portion of the gay community can no more choose to be attracted to the opposite sex as those who are straight can choose to be attracted to those of the same sex. This means that we also need to factor in the moral problem with refusing to accommodate a significant number of people because they are different.

Uncomfortable Idea: Just because male homosexual sex is riskier, does not make it morally wrong, just like riding a motorcycle is no more morally wrong than driving a car.

Remember that "gay" refers to same-sex attraction, behavior, or identity. People can be convinced through a therapeutic process to abstain from same-sex behaviors, and they can be convinced to identify as straight. Bisexuals can be convinced to only act on their heterosexual desires. **In no way, however, does this mean that gays can choose to be attracted to the opposite sex, or that this kind of therapy is a good idea.** The official statement from the American Psychological Association (APA) on this issue reads

> *All major national mental health organizations have officially expressed concerns about therapies promoted to modify sexual orientation. To date, there has been no scientifically adequate research to show that therapy aimed at changing sexual orientation (sometimes called reparative or conversion therapy) is safe or effective. Furthermore, it seems likely that the promotion of change therapies reinforces stereotypes and contributes to a negative climate for lesbian, gay and bisexual persons. This appears to be especially likely for lesbian, gay, and bisexual individuals who grow up in more conservative religious settings.*[64]

Most people and organizations who are against normalizing homosexuality incorrectly believe that gays can be "cured" of their same-sex attraction. They use anecdotal evidence in an attempt to support these claims. Gays who claim they have been cured of their gayness have powerful biases for making those claims that result from fear of eternal damnation, expulsion from their religious communities, being kicked out of their homes, being shunned by family members and friends, divorce, losing their children in a custody battle, getting fired, and other legitimate fears that result from being gay—especially in a conservative family, community, or part of the country. This is why scientific methodology and research is far superior to "I know this guy who..." stories.

---

[64] Answers to Your Questions For a Better Understanding of Sexual Orientation and Homosexuality. (n.d.). Retrieved October 7, 2016, from http://www.apa.org/topics/lgbt/orientation.aspx

Uncomfortable Idea: Gay conversion therapy can make gays no longer identify as gay and make them not want to engage in same-sex behavior.

Fantasy is Sometimes Better Than Reality

Nature has not made it a priority that we know the truth—not when the truth interferes with survival or reproduction. As mentioned, the self-serving bias is our brain's way of interpreting information that makes us feel good about ourselves despite the fact that the reasons why we feel good about ourselves, or life in general, are grossly inaccurate. People who believe that they will spend eternity with their loved ones when they die do not want to hear evidence against such a claim; thus the idea is avoided. The self-serving bias is just one of the several biases that protect us by favoring a comfortable lie over an uncomfortable truth.

Civilization as we know it would collapse without people doing those crap jobs that most of us don't want to do. Some of us have greater intellectual abilities, more ambition, more motivation, and more resources to make use of the natural abilities we do have. Others are incapable of curing cancer, being an astronaut, becoming an all-star basketball player, winning an academy award, or saving the world. The fact that any given person is overwhelmingly in the latter group does not mean we shouldn't try for greatness; it just means that we shouldn't count on it, be disappointed when our children don't live up to our unrealistic expectations, and belittle or undervalue those who are doing what we consider the crap jobs that keep the world spinning.

Uncomfortable Idea: Not everyone gets to be an astronaut.

The whole "pick yourself up by the bootstraps" myth is a destructive one that is a major insult to billions of people based on sheer ignorance. This idea stems from the *just-world fallacy*, which states that people believe in a just world where everyone gets what they deserve. Believing in a just world greatly reduces the pain one might experience in knowing that people who suffer are suffering because they somehow deserve it. Or for those who extend justice to other lives, justice will be sorted out in the afterlife—so it's not a big deal that good people suffer... in fact, it's good for them.

Uncomfortable Idea: The world is not just. Sometimes the bad guys win, and sometimes the good guys lose.

We can cite data and refer to statistics that show without question that financial success is largely a result of the socioeconomic climate in which a person was raised, but this wouldn't matter. Just-world folks like to use anecdotes in place of statistics, confusing possibility with overwhelming probability. They like to cite "American success stories" or rags-to-riches tales of people like Oprah Winfrey, a poor black girl raised in poverty and sexually abused as a child who became one of the world's wealthiest people. If Oprah can do it under those circumstances why can't every else? Besides Oprah having superior intellectual ability, natural talent, and natural beauty, she had the motivation, ambition, perseverance, and opportunity. Even if we assume that all or even most people living in poverty have enough intellectual ability and/or talent to be financially successful (which is a poor assumption based on fantasy, not reality), should we hold them responsible for their lack of motivation, ambition, perseverance, and other traits strongly correlated with financial success? If we ignore the genetic component to these traits, we only have environmental factors left. Clearly, we can't blame people for their genetic make-up, but can we blame them for their environment? How many children raised in poverty had positive role models who provided the encouragement needed to succeed financially? Now add in bad luck, and it should be easy to see that very seldom are people are to blame for their lot in life. But we are to blame for not doing more to help them.

Uncomfortable Idea: Poor people are rarely to blame for being poor, and rich people get far too much credit for being rich.

Obese people are not perfect the way they are. This isn't some Hollywood-manufactured obsession with being thin conspiracy. Obesity-related conditions include heart disease, stroke, type 2 diabetes, and certain types of cancer, which are some of the leading causes of preventable death.[65] Perhaps you think that you can do whatever you want to your body as long as it doesn't hurt others. You might be the type who doesn't like helmet or seatbelt laws. That's fine, but that still doesn't make you perfect. If you want to justify and rationalize your obesity, go ahead. Without question, genetics, medical conditions, and other factors have a significant impact on one's likeliness of being obese. Just don't glorify obesity and attempt to make it the new normal for the sake of the rest of humanity who have greater control over their general fitness.

Uncomfortable Idea: Obese people are not perfect the way they are. Improved health and well-being might have to come at the expense of protecting obese people's sensibilities.

The Work of Satan

Yes, there are still an alarming number of people who genuinely believe that entertaining certain ideas is "the work of Satan." In the Bible, the Satan character is known for his deception of man. This gives believing Christians an easy out for any idea that contradicts their beliefs. Your friend who is asking you why having faith is a good idea is being tempted by Satan. That book that looks interesting on biological evolution and natural selection was inspired by Satan. That

---

[65] Adult Obesity Facts | Overweight & Obesity | CDC. (n.d.). Retrieved from https://www.cdc.gov/obesity/data/adult.html

church across the street that is welcoming to gays has been taken over by Satan! Satan, the great deceiver who even disguises himself as an angel of light (2 Cor 11:14) apparently only fools other people, so there's nothing to worry about.

It's clear why people would not want to be influenced by Satan. Satan, after all, is the most horrible character imaginable. He has commanded humans to kill gays, kill kids who disrespect their parents, kill adulterers, kill non-believers, kill women who are not virgins on their wedding night, kill people for working on the Sabbath, and kill people who don't believe as you do. Whoops. Scratch that. That was all God.

Uncomfortable Idea: God is a far more horrifying character than Satan.

Fear of Entertaining Evil, Sick, or Immoral Thoughts

Social psychologist, Jonathan Haidt, poses the following scenario:

A brother and sister are vacationing in the south of France. They have some wine, one thing leads to another, and they decide they want to have sex. They use two different kinds of contraception and enjoy it, but they decide not to do it again. Is there anything morally wrong with this?

If you are like most people Haidt surveyed, you will immediately have feelings of disgust and immediately object to this scenario. When Haidt's subjects were asked for reasons why they object, none were offered besides "it's just wrong" or some similar reaction that avoids explaining why.[66] We avoid entertaining ideas that just feel wrong simply because the ideas to us are automatically interpreted as evil, sick, or immoral. The ironic part is, unless we entertain these thoughts, we cannot possibly justify them being evil, sick, or immoral.

---

[66] Haidt, J. (2013). *The Righteous Mind: Why Good People Are Divided by Politics and Religion* (Reprint edition). New York: Vintage.

Uncomfortable Idea: Sometimes, some really disturbing ideas are difficult if not impossible to justify as morally wrong.

In the United States, the age of consent ranges from 16 to 18, which means that it is legally wrong for anyone under this age to engage in sexual activity. Worldwide, this age ranges from 11 to 21. These laws are mostly to keep adults from taking advantage of kids sexually but still applies to two consenting youths under this age. Cultural and social norms combined with moral views dictate local laws, and local laws often affect moral views. It's important to recognize that laws such as these apply to behavior, not feelings.

Physical attraction is an automatic and feeling-based process, whereas *behavior*, such as engaging in sexual activity, is the result of deliberate cognition. We may feel at times that we want to kill our boss, but we don't, because we are capable of higher-order thinking that can override our primitive brain's desires and impulses. We can control our behavior, but not our attraction. Most people might agree that feeling like killing one's boss is normal, as long as that feeling is not acted upon or seriously considered, but what about being attracted to a sexually-developed adolescent under the legal age of consent?

Visible signs of puberty start appearing around age 10–11 for females and 11–12 for males, and puberty is completed around age 15–17 for females and 16–17 for males. Attraction is a biological process that is necessary for the survival of our species. Evolution does not care about age of consent laws, so attraction to sexually-developed humans of the opposite sex is a normal, healthy, feeling no matter what the age. Men are more likely to experience this physical attraction than women for reasons that can be traced back to our evolutionary history. Since women bear the cost of pregnancy and lactation, their concern was with acquiring resources to produce and support offspring. Men faced the adaptive problems of identifying fertile partners of high reproductive value. In other words, when it comes to attraction, physical features are more important to men while stability and security are more important to women. Since teenagers rarely display stability and security but commonly stand out because of their physical features and youth, it's far more common that older males will find younger females attractive than older females finding younger males attractive.

Pedophilia is not a term that should be taken lightly and watered down by applying it to all men who happen to find a sexually-mature 17-year-old girl attractive. *Pedophilia* is a serious psychiatric disorder in which an older adult is primarily attracted to **pre-pubescent** children. It becomes a crime when people with pedophilia act on their attraction. The terms *hebephilia* and *ephebophilia* refer to older adults who are primarily attracted to pubescent (ages 11–14) and mid-to-late adolescents (ages 15–19) respectively.

This is one of those dangerous truths that might benefit society by remaining a taboo idea. Perhaps it's better that we all keep pretending that everyone under the legal age of consent is not at all sexually attractive, but a day later when they reach that legal age, they're smokin' hot. Without the taboo, acting on attraction becomes more likely. Feelings lead to cognitions, and cognitions lead to behaviors. Subtle behaviors can be inappropriate displays of attraction such as flirting or inappropriate comments among friends. These subtle behaviors can turn a harmless attraction into an obsession and ultimately result in harmful illegal sexual behavior. Finding a teenager attractive may not make you perverted, sick, gross, in need of counseling, or a pedophile, but acting on that attraction might just land you in prison.

Uncomfortable Idea: Finding a sexually-mature teenager attractive doesn't make you perverted, sick, gross, in need of counseling, or a pedophile.

There are over seven billion people in the world, and unless one of those people is the Buddha, they all have needs and desires. In the United States alone, there are over 1.5 million non-profit organizations,[67] most of which are trying to convince us that their particular cause is of utmost importance. A good number of them are probably using that clichéd line "for the cost of a single cup of coffee each day..." Since we don't drink 1.5 million cups of coffee a day (i.e., you or I don't have about 4.5 million dollars a day to spend), and because our resources are limited, we cannot possibly devote our resources (including emotional ones) to every issue that catches our attention, whether it be Sally Struthers on the television or a homeless man on the street. That's okay. Don't beat yourself up for walking past the homeless, hanging up the phone on solicitors for charities, and not bidding $1000 at the charity auction for the blender that retails for $39.95. Be generous, be kind, and be realistic. You can't solve all the world's problems so don't feel guilty for not trying.

Uncomfortable Idea: It's okay not to care about everything and everyone, as long as you care about something and someone.

Fear of Questioning / Refusal To Question Authority

Some refuse to entertain ideas that go against an authority figure in fear that they will be punished for questioning authority, displaying a lack of loyalty, showing a lack of faith in the authority, or simply having doubts. Even if punishment is not a concern, guilt can be. "Authority" can be a god, religious or cult leader, teacher, parent, boss, a document that one holds sacred like the Bible or Constitution, current laws, rules, or even one's "gut."

In Islam, those who leave the faith, can be killed for their actions. This is a pretty strong motivator not to entertain the idea that the religion is false. In Christianity, according to 2 Thessalonians 1:8, punishment for not believing in God is being burned to death by Jesus and his angels. Again, if you believe the authority of the Bible, this is a pretty strong motivator not to entertain ideas that might lead you to no longer believe—just in case you might be wrong.

---

[67] Quick Facts About Nonprofits. (n.d.). Retrieved October 9, 2016, from http://nccs.urban.org/statistics/quickfacts.cfm

Uncomfortable Idea: Getting killed for not believing in Allah or believing that you will be tortured for eternity for not accepting Jesus as your lord and savior are extremely strong motivators for avoiding arguments against your belief system.

A refusal to question authority can also be a result of complete trust or "faith" rather than fear. Thinking is hard. Moral dilemmas are challenging, as is making decisions where disagreement, pain, and suffering will definitely result. To avoid this responsibility and the associated stressors, we can put our complete trust or faith in another person or idea. When we do this, we trade our morality for obedience. If we want to know if an action is right or wrong, we don't use our moral reasoning and think about the consequences of the action; we simply defer to the authority figure. Once we question this authority figure, we hop on the slippery slope that can put us back in the moral driver seat of our own lives where we are once again responsible for our decisions and our lives. Some people cannot handle this kind of stress, or simply do not want it. This is why many people do not entertain ideas that go against their authority figures' ideas; they simply defer to their authority figures' views on the idea—or use their own mortality to interpret their authority figures' views on the idea in such a way that matches their views. This is a win-win because they're still calling the moral shots but with none of the responsibility.

Uncomfortable Idea: Choosing to be obedient over considering the short- and long-term consequences of your actions and how they affect people is a moral cop out.

Fear of Confusing Support for Personal Desire

I have to admit, writing the section on polyamory, polyandry, and polygamy made me uncomfortable (much of this book did, in fact). As a social psychologist and a self-proclaimed moral philosopher, I am open to entertaining these ideas, specifically, looking at the data and seeing how these kinds of relationships affect the partner's well-being and the well-being of family members. I am sure a sociologist would have other concerns such as how allowing and supporting these relationships would affect society in general. However, I have no personal desire to take another wife or start sleeping with other people, nor do I want my wife to take another husband or start sleeping with other people. I don't actively support polyamory because I don't care enough about it to spend my time on the issue, but I also refrain from discussing it because I don't want others to get the impression that I am looking for that kind of relationship. Let's look at another idea that would make any faithful spouse cringe.

One can truly love their spouse and still cheat on them. Again, it's difficult to imagine any married person promoting this idea without having a not-so-hidden agenda. The consequences for supporting (perhaps even entertaining such an idea, depending on the spouse) can be disastrous for the supporter's marriage. No matter what the data say, because perspective trumps data, accepting the idea that one can truly love their spouse and still cheat on them is akin to asking permission to cheat and giving permission to one's spouse to cheat. Let's look at a couple of other issues.

Uncomfortable Idea: One can truly love their spouse and still cheat on them.

I support legalized marijuana. Despite this fact, I have never used the drug, never wanted to, never plan to, and would strongly encourage others to stay away from it. Although I do think this is an important issue, I rarely discuss it or write about it, partly because I don't want people thinking I am a stoner. Similarly, there are many people who refuse to accept ideas because **accepting an idea on ideological grounds can be misinterpreted by others as accepting the idea for personal reasons**. Legalized prostitution is another less socially acceptable idea that is easy to understand why politicians wouldn't want to support. The fear of being misunderstood and the associated consequences can be so great that these kinds of issues are publicly dismissed and unfairly demonized, despite the possible social good they may offer.

Fear of Exposing Our Own Demons

We all have idealized ideas of who we want to be as individuals, and what we should be as a human race. But like all concepts of perfection, they don't exist in the real world as they forever remain just out of reach. Those thoughts, feelings, or behaviors that are not consistent with the kind of person we think we should be are our metaphorical demons. We don't like ideas that remind us of how we fail to live up to our own standards, so we avoid entertaining them.

We can be a supporter of gay rights, a gay rights activist, and still be disgusted when seeing two men make out with each other. Visceral feelings of disgust don't disappear just because you approve of that which makes you disgusted since the reaction is an automatic one and does not involve cognitive evaluation. This reaction could be explained in part by *mirror neurons*, brain cells that trick our brain into thinking that we are experiencing what we are watching another person experience. And for those of us who aren't gay, trading spit with a person of the same sex could be a very unpleasant experience. We might **want** to be comfortable watching gays kiss and we might even feel guilty for not being comfortable with it, but this doesn't make us anti-gay or homophobic.

Uncomfortable Idea: Being disgusted by watching two people of the same sex passionately kiss does not make you homophobic.

Who you find attractive or not is mostly an unconscious process that does not shut off simply because you're wearing a ring on your left ring finger. There is no evidence to suggest that true love has any effect on the ability to find other people attractive. Fantasizing about others, flirting with others, and being intimate with others who are not your spouse are all very different from attraction, as we have control over those behaviors (less with fantasizing than with the others). There is also no evidence to suggest any correlation with finding other people attractive and the quality of one's relationship with their spouse.

Uncomfortable Idea: It's not a moral failing to be attracted to people who aren't your significant other.

"Inappropriate" jokes are funny, and there's no better way to expose our inner demons than through humor. If we come across some form of satire that exaggerates stereotypes sometimes to the point of absurdity we laugh—even though we might find the satire tasteless, crude, and offensive. Laughter is not within our conscious control.[68] A laugh can be stopped or held back consciously (like when your friend farts at a funeral), but the laughter begins as an automatic and unconscious reaction to stimuli. Consider this joke from the movie *Guess Who?* starring Ashton Kutcher and Bernie Mac:

What are three things a black guy can't get? A black eye, a fat lip, and a job.

How did you feel when you read this? You might have experienced an initial positive emotion followed quickly by its suppression, then perhaps a lingering feeling of guilt. Are you a closet racist? Probably not. You most likely have a developed sense of irony and reacted to the "list of three" rule of humor where the third item on the list was unexpected due to *equivocation*, or changing the use of the term "get." Once our conscious evaluation kicks in, we realize that unemployment in the black community is a serious social issue and being reminded of such a problem quickly counteracts any positive emotion. It's not the social injustice that we find funny; it's the irony in the joke itself. Finding inappropriate jokes funny is very different from spreading inappropriate jokes (writes the guy who just spread an inappropriate joke).

---

[68] Provine, R. R. (2001). *Laughter: A scientific investigation*. Penguin.

Uncomfortable Idea: It's okay to find inappropriate jokes funny.

Upon closer inspection, we will find that our "demons" are really just misunderstandings. They are reflections of our humanity, or more accurately, results of a normally-functioning mind. The better we understand this, the less likely we are to avoid these kind of ideas that make us feel guilty for being human.

Part III: Why We Refuse To Accept Uncomfortable Ideas

A mother gets a call from the local police department. Her 16-year-old son has been arrested for driving under the influence. Her first reaction is "this must be some kind of mistake." After the police confirm that they have the right person by describing the son and reading the details of his ID, the mother goes to the police station and demands to see the results of the field sobriety test. When the mother is informed that her son refused to take the test, she is relieved and concludes that this is some elaborate setup—as she helps her clearly inebriated son (clear to everyone but the mom) from the police station into her car.

When a person refuses to **entertain** an uncomfortable idea, they shield themselves from evidence and facts that could cause them to accept the idea. When the idea is already entertained, and the evidence and facts would make accepting the uncomfortable idea the only rational choice, the decision-maker has two irrational options in addition to the rational one: 1) refuse to accept the evidence and the facts or 2) refuse to accept the uncomfortable idea itself (the conclusion).

Some of the reasons we refuse to accept uncomfortable ideas or the evidence and facts that support them are the same as why we refuse to entertain them. For example, we might value feeling over fact. All the facts can point to the mother's son driving drunk, but it simply does not "feel" right. These kinds of feelings are known as *intuition* and are based on our experiences, both conscious and unconscious. Intuition is not reliable in this case since her son's behavior is unlikely to be consistent at age 16. In other words, her experience with the five-year-old version of her son not driving drunk only misleads the mother to having false confidence in her son. The mother could have also made the belief sacred that her son would never do such a thing as drive drunk, thus dismissing the facts for that reason alone. Or perhaps the mother might accept that her son was caught driving drunk, but refused to accept that it was his fault because he was manipulated by Satan at the time.

In this section, we will look at some cognitive biases that might explain why otherwise reasonable people refuse to accept uncomfortable ideas or the facts and evidence that support the ideas. But first, let's take a quick look at how we evaluate evidence.

Evaluating Evidence

Consider this statement from Francis Collins, former head of the Human Genome Project and Evangelical Christian:

As someone who's had the privilege of leading the human genome project, I've had the opportunity to study our own DNA instruction book at a level of detail that was never really possible before. It's also now been possible to compare our DNA with that of many other species. The evidence supporting the idea that all living things are descended from a common ancestor is truly overwhelming. I would not necessarily wish that to be so, as a Bible-believing Christian. But it is so. It does not serve faith well to try to deny that.[69]

Here is a guy who understands DNA better than just about anyone on the planet, as well as understands the implications that all living things definitely did not just appear on the earth in their current form. Yet as a self-proclaimed Bible-believing Christian, his understanding has turned his worldview upside down and caused him to rethink how he understands his faith. Yet according to a 2014 Gallup poll, 42% of Americans still believe that a god created humans in present form. It's not like 42% of Americans have never heard of DNA. Those who haven't, or don't understand the implications, are probably aware of the fossil evidence or at least they are aware of the virtually universal scientific consensus that exists among scientists that life evolved (97% agreement[70]). So how can people interpret the same information so

---

[69] "God Is Not Threatened by Our Scientific Adventures." (n.d.). Retrieved from http://www.beliefnet.com/news/science-religion/2006/08/god-is-not-threatened-by-our-scientific-adventures.aspx

[70] Street, 1615 L., NW, Washington, S. 800, & Inquiries, D. 20036 202 419 4300 | M. 202 419 4349 | F. 202 419 4372 | M. (2009, July 9). Section 5: Evolution, Climate Change and Other Issues. Retrieved from http://www.people-press.org/2009/07/09/section-5-evolution-climate-change-and-other-issues/

differently? Remember ABC—awareness, believability, and comprehension.

Awareness

It should go without saying that one cannot evaluate evidence without being aware of the evidence. Avoiding uncomfortable ideas is the same as avoiding evidence that would support such ideas. We can avoid evidence ourselves or be under the control of a gatekeeper who keeps evidence from us, or presents a *strawman* version of the evidence (i.e., an inaccurate or poorly presented version of the evidence). This is common among creationists who homeschool their children and keep them away from the science of evolution.

With most issues of science, the general public is simply not aware of the extent of the research done. The public may come across a sexy headline that reads "A Glass of Wine is Better Than an Hour at the Gym" and be sold on the idea that drinking booze is better than working out while being blissfully unaware of the collection of evidence that contradicts that headline written to get clicks. The general public is exposed to what people in the media think the public would be interested in knowing. The rest of the information remains buried in obscure science journals.

We need to be made aware of the evidence—not some inaccurate version of the evidence presented in such a way so any reasonable person would dismiss it, but the facts presented in an unbiased way. Then we can move on to looking at the believability of the evidence.

Believability

Believability in this context refers to knowing what level of trust one should have in the source as well as knowing what constitutes evidence and how to tell the difference between strong and weak evidence. While a discussion of the different types of evidence are beyond the scope of this book, it is worth listing some different types of evidence and the general order[71] based on strength (strongest to weakest):

---

[71] I say "general order" because each type of evidence has its own level of strength. For example, finding a bloody glove at a crime scene with the accused's DNA on it, is strong physical evidence and stronger than say a poll asking people who they think "did it."

- Systematic review, meta-analyses, strong scientific consensus
- Randomized controlled double-blinded studies
- Cohort studies
- Statistical evidence (e.g. surveys, experiments, polls)
- Physical evidence (e.g., a bloody glove found at a crime scene)
- Case studies
- Scientific / expert opinions
- Testimonials / eyewitness testimony
- Anecdotes

Notice that "personal feelings" is not on this list. If a man is on trial for murder, a testimony of "I just know he did it—I feel it" doesn't go very far. Scientists who bring new medicines to market don't tell the FDA "I just feel like this drug would work." You are the authority on how you feel, but not on objective truths about the world that you infer based on how you feel or perceive the world around you. This brings us to the problem of overreaching.

Overreaching is when a person adds unwarranted and unsubstantiated details to a subjective experience. For example, if you see a flashing light in the sky and claim that is evidence for alien spaceships, you're overreaching. If you are overcome by a feeling of peace and claim that is evidence for Jesus, you're overreaching. If you have a dream about a friend who happens to call you the following day and you claim that is evidence for ESP, you're overreaching. What we actually have is evidence for blinking lights in the sky, feelings of peace, and events that seem strange and improbable to us.

Once we understand the nature of evidence, we can look at the specific form of evidence and see how reliable that evidence is. While it is true that systematic reviews are generally more reliable than anecdotes, each specific form of evidence has its own scale of reliability based on the source. For example, roughly 3% of scientists are creationists—people who have adjusted their interpretation of scientific data to be consistent with their religious views. If we look at scientific/expert opinion, we can have one scientist who agrees with the consensus of evolution, and another who believes that man was created in his present form. How we evaluate those expert opinions will be based on how much we trust each scientist who is offering the opinion. What makes us trust one more than the other? Our biology and environment—pretty much how we are wired combined with every life experience we had.

Before we can evaluate evidence, we need to know what evidence is, how to tell the difference between weak and strong evidence and understand the concept of overreaching. We also need to evaluate the level of trust we should be giving to the source. Then we can move on to looking at how we comprehend the evidence.

Comprehension

It may not be unreasonable to bypass the comprehension of evidence because of your trust in the source. Of course, your trust in the source needs to be reasonably justified. If your doctor tells you that you have strep throat, it would be reasonable to accept her conclusion given the relative consequences of her being wrong and not believing her if she is right. You don't need to crack open the medical books to learn all about the infection and run the lab tests yourself. Your doctor's expert opinion is good enough in this case. However, in most cases, you want to seek comprehension. The more you generally understand, the better your conclusion.

Comprehension, or understanding the evidence is helpful when evaluating it because simple fiction often wins when up against complicated facts. If we have a look at the hundreds of creation myths found in different cultures, we will find that all of them are simple and easy to understand, usually because of the use of some form of magic that does not need to be explained. Comprehending scientific facts about our universe often requires a good understanding of the science in general, and in many cases, a strong multidisciplinary understanding.

George Burns once quipped, "Too bad that all the people who know how to run the country are busy driving taxicabs and cutting hair." If he were around today, he might have said, "Too bad that all the people who know how to run the country are busy posting on Facebook and Twitter." Common folk evaluate evidence differently than those who are privy to all the details. We tend to criticize our leaders based on a cursory understanding of the issues at best, not acknowledging that they have a much more comprehensive understanding of the issues than we ever could, given their access to information that we don't have. Lack of comprehension is bad enough, but when that ignorance is combined with the illusion of understanding, our evaluations of evidence can be grossly inaccurate.

At this point, we have a sense of why some people evaluate evidence differently than we do, which can lead them to not accepting an uncomfortable idea that should be accepted if they were made aware of the evidence presented in a fair and honest way, if they understood the nature of evidence and the credibility of the source, and if they were able to comprehend the evidence and its implications reasonably. Now let's have a look at some specific ways in which our brains prevent us from accepting uncomfortable ideas, even when we can't reject the evidence.

Belief-Related Cognitive Biases and Effects

At the beginning of this section, the example of creationism was used to demonstrate how people might come to different conclusions given the same evidence. In the section on evaluating evidence, it was explained in more detail how awareness, believability, and comprehension might account for how we evaluate evidence. But can people accept the evidence that supports an uncomfortable idea yet still reject the idea itself? Technically speaking, accepting the evidence is not the same as not rejecting the evidence. This is where feeling over fact comes into play. We could have no reason for rejecting the evidence, yet at the same time, we don't have to accept it. We see this often with creationists who flat out admit that no matter what the evidence shows, "God said" he created man in the present form, so that is that. Another example is our mother who "just knows" her son is innocent of drunk driving. Thanks to belief-related cognitive biases and effects, sometimes facts, evidence, and reason don't stand a chance.

Backfire Effect

Our emotional attachment to an idea can be so strong at times that our level of irrationality borders on humorous. The *backfire effect* occurs when we are presented with compelling evidence against our position, and instead of accepting the evidence, we push back even harder with more arguments, often based on exaggeration, wishful thinking, and outright lies. If you get into a debate about climate change with a friend who does not accept that humans are responsible, you might find that the more data, science, and facts you throw his way, the more convinced he seems that his position is right. Instead of refuting your facts, your friend gets deeper and deeper into the conspiratorial nature of climate change and how all the scientists are on someone's payroll. If you're lucky, you may even get to hear about the Illuminati.

Belief Bias

In this context, the *belief bias* is the tendency to base the acceptance of an idea on the plausibility of the idea rather than how strongly the evidence supports the conclusion. Think of a husband who could never even imagine that his wife would ever cheat on him. To the husband, the idea of having a cheating wife is extremely implausible. Therefore, the uncomfortable idea that his wife is actually cheating on him is rejected, even though the evidence clearly supports the cheating wife theory. The husband won't deny the facts; he will just reject the obvious implications (obvious to everyone else). The fact that his wife gave birth to a black baby, even though the husband and his wife are white, must be due to some genetic anomaly because his wife would never cheat on him.

Confirmation Bias

In this context, the confirmation bias is used to interpret evidence in such a way that supports the comfortable idea. Unlike the belief bias, the evidence isn't necessarily dismissed; rather it is interpreted in a way that supports the desired conclusion. Sticking with our cheating wife example, imagine the husband confronting his wife about a room she got at a sleazy motel a few days earlier. The wife says she was planning a sexy mid-day getaway for her and her husband, and just wanted to make sure the motel was clean. Thanks to the confirmation bias and the strong desire to not accept the uncomfortable idea, the husband would see this as evidence **against** the idea that his wife is cheating on him rather than evidence supporting the idea.

Ostrich Effect

When applied to uncomfortable ideas, the ostrich effect is the belief that an uncomfortable idea will go away if you ignore it. You might have no immediate objections to the veracity of the idea or to the facts supporting it (but still dislike it emotionally), but because the idea is a fringe one or it is the first time you hear it, you ignore it in hopes that it will go away and you won't have to consider it again. To give you a personal example, I have been avoiding getting a flu shot— to me the need for getting a flu shot is an uncomfortable idea. I cannot argue with the science or the reasons for getting the shot, but I didn't accept the idea either. I ignored it... until I got the flu.

Status Quo Bias

According to the *status quo bias*, people prefer things the way they are. If we are presented convincing evidence for an uncomfortable idea, but the evidence is not significantly stronger than the evidence we have for our comfortable idea, we are likely to stick with accepting our comfortable idea. In other words, for us to change our position on something the evidence needs to be more compelling than if we had no opinion on the issue to start with. If you hold the belief that prayer works (in that a god will respond favorably to your desires), and you discover that although this idea has been thoroughly tested over decades and no good evidence has been found to support such a claim, the fact that no evidence has been found to **disprove** the claim might be enough for you to hold onto your belief, or maintain the status quo.

There are literally hundreds of biases, fallacies, heuristics, and effects that affect our beliefs. These are just a few that most likely play a role in causing us not to accept uncomfortable ideas, regardless of the quality of the evidence.

Refusal to Accept Due to Refusal to Reject

We need to remember that there are often two sides to the equation when it comes to accepting an uncomfortable idea: there is the acceptance of the uncomfortable idea and the rejection of the comfortable idea that it will replace. If we were to accept the uncomfortable idea that there is no afterlife and when we die we just cease to exist, to be logically consistent we would also need to reject the comfortable idea that we are immortal and our eternity will be nothing but perfection. We shouldn't indiscriminately reject ideas just because they are comfortable any more than we should reject ideas just because they are uncomfortable, but if we are to evaluate uncomfortable ideas honestly, we need also to reevaluate competing ideas that may initially bring us more comfort.

If you are evaluating the idea of evolution, then learning everything about it won't be enough. You will need to reevaluate your competing beliefs that might include the idea that the creation story in the Bible is meant to be taken literally, that Satan is trying to deceive you, or that accepting evolution will cause you to become an atheist. If you are evaluating the idea of human-caused climate change, you will need to revaluate your ideas of global conspiracies and the validity of politician's views on climate change rather than climatologist's views on climate change. And if you are convinced that your wife would never cheat on you, you may need to consider reevaluating the ideas that your wife is perfect, your ability to satisfy her is top-notch, and that Rafael, your wife's 22-year-old administrative assistant, is gay.

We don't process information in a vacuum. Our beliefs are all connected in order to hold consistent worldviews. One simple uncomfortable idea such as "God exists" for the atheist or "God does not exist" for the theist, can lead to a chain reaction that compromises hundreds or even thousands of beliefs that are dependent on the foundational belief. At some level, we understand this and protect against this kind of epistemological crisis. This is why those ideas that challenge our foundational beliefs are the most uncomfortable ones and the most difficult to accept.

Part IV: Some More Uncomfortable Ideas

I've attempted to write this book to be an "equal opportunity offender" in that no matter what your religious affiliation, political leaning, sexual orientation, gender, race, or favorite ice cream flavor, reading some of these ideas should make you uncomfortable or even make you furious. If you haven't felt this way yet, at least some of the ideas in this section should do the trick.

The Self-Fulfilling Nature of Social Injustice

There are millions of Americans who rely on "good luck charms" to help them through life. People use rabbit's feet, four-leaf clovers, horseshoes, and many other objects for the purpose of attracting good fortune in their lives. Even professional athletes often have good-luck rituals that they perform before each game or event. If these knick knacks and rituals don't really have any magical powers, why do they appear to work as made evident by the millions of Americans still using them? The answer is what is known in psychology as the *self-fulfilling prophecy*, which is an assumption or prediction that purely as a result of having been made, cause the expected or predicted event to occur and thus confirms its own "accuracy." [72] In other words, if we believe that we will have good luck, through what is known as the *frequency illusion*, we will notice good luck everywhere. Through the confirmation bias, we will tend to remember only the good luck and ignore or forget about the bad luck, thus reinforcing and strengthening our belief in our good luck charm or ritual. Unlike the professional baseball player who wears the same unwashed underwear to each game, the social justice activist doesn't require a charm or ritual, but does create social injustice by how they interpret events.

Social justice is commonly defined as "justice in terms of the distribution of wealth, opportunities, and privileges within a society." This is a good thing, and something everyone should stand for. The problem is, "justice" is a very subjective term where people have very different ideas as to what constitutes justice. This is why there are

---

[72] Watzlawick, P. (1984). Self-fulfilling prophecies. *The Production of Reality: Essays and Readings on Social Interaction*, 392–408.

countless organizations that focus on their idea of "justice," from Black Lives Matter activists who believe "Black lives are systematically and intentionally targeted for demise"[73] to The Knights Party who believe that within 25 years white people will be "a minority that will be in the midst of a genocide."[74] There are groups that fight for and against vaccines, anti-abortion, pro women's rights, groups for the death penalty and against, groups who want more God in government and groups that want no God in government. No matter which of these groups are right, they all strongly believe they are doing the right thing, that is, they all believe that they are warriors for justice.

Some of the most effective activists are ones who are more like lawyers than scientists; they will bend facts, twist the truth, cherry pick data, sell fear by sharing the most pessimistic forecasts, and outright lie when needed. Facts and data are no match for strong fallacy-ridden emotional appeals. In this sense, activism is antithetical to critical thinking, reason, and science. There is an old saying that we are the heroes of our own story, which means that most of us think we are saving the world in our own little way. Heck, I believe I'm doing my share by writing this book. Thanks to the self-serving bias, we can strongly believe this because it contributes to our well-being. It doesn't matter if we are completely deluding ourselves and nature doesn't care —as long as our delusions don't get in the way of our survival and reproduction.

To be a hero, we need a villain. The more evil the villain, the greater the hero. In the world of social justice, the villain is the social problem. The people who are said to be part of the problem, the people who don't agree that the problem exists, and even the people who don't actively do something about the alleged problem, are also the villains. It is human nature to vilify and demonize, ignore the data and arguments that conflict with our point of view and create ingroups of

---

[73] Guiding Principles | Black Lives Matter. (n.d.). Retrieved from http://blacklivesmatter.com/guiding-principles/

[74] The Knights Party FAQ. (n.d.). Retrieved September 21, 2016, from http://kkk.bz/?page_id=2896

heroes and outgroups of villains. As a side effect of these behaviors, we are intensifying and in some cases even creating social injustice.

Take any social issue. How common or *pervasive* is the problem? Remember, nobody's interested in a villain whose just kind of a dick and sometimes does bad things. We want to battle a villain so evil that he is everywhere doing horrible things all the time. The solution? Assume the least charitable interpretation that intensifies or creates the villain. **By interpreting an event in a racist, sexist, homophobic, xenophobic, or otherwise intolerant way where no such intolerance was intended, we are creating our villain and creating social injustice.** A prime example is the previously mentioned photoshopped Tweet of Ellen DeGeneres getting a piggyback ride from Usain Bolt.[75] It is extremely likely that Ellen, arguably one of the sweetest people in show business, had no racist intent. Social justice activists had to dig deep to find racism by relating slavery to treatment of animals, and because people ride animals, then Ellen must be racist. **The strong belief that racism is everywhere creates the reality that racism is everywhere through the self-fulfilling prophecy.** Also, by trivializing racism in this way we are doing a disservice to the real injustice that is racism. Common sense should tell us that sticking Ellen Degeneres in the same category as David Duke, former leader of the KKK, is just wrong.

Another example of the self-fulfilling nature of social injustice can be found with the famous *V-J Day in Times Square* photo showing a sailor kissing a nurse. What was once a photo depicting victory, celebration, and the spontaneity of romance, is now being repackaged as "an act of sexual assault."[76] When the image was originally published in *Life* magazine the caption read "In New York's Times Square a white-clad girl clutches her purse and skirt as an uninhibited sailor plants his lips squarely on hers." Although the act captured on film did not change since 1945, the interpretation did. Due to this

---

[75] Mic. (2016, August 16). The "Ellen Show" Tweeted a Pic of DeGeneres Riding Usain Bolt's Back Like an Animal. Retrieved from https://mic.com/articles/151655/the-ellen-show-tweeted-about-usain-bolt-and-people-are-calling-it-racist

[76] Iconic kissing sailor photo depicts sexual assault, not romance. (n.d.). Retrieved September 21, 2016, from http://feministing.com/2012/10/04/iconic-kissing-sailor-photo-depicts-sexual-assault-not-romance/

change, the world has one more act of "glorified sexual assault promoting rape culture" and one less spontaneous act of romance.

Let's not be guilty of romanticizing; there is, by all reasonable standards, much injustice in the world and we should all be grateful for those who stand up to this injustice. But this injustice should be fought with honesty, reason, and integrity. If you are one of the many people fighting for injustice, ask yourself how much of what you do has to do with making yourself a hero? How confident are you that you are on the right side of the battle? Are you being charitable or even just reasonable in your interpretation of events, or are you viewing the world through intolerance-colored goggles? Don't be like Sigmund Freud and interpret every thought and behavior as penis envy. Sometimes a joke is just a joke, a kiss is just a kiss, and a cigar is just a cigar.

Love Isn't Always Beautiful, and You Don't Love Everyone

If you are a parent, imagine yourself in a situation where you have to make the decision to save either your child or a dozen children who you don't know. What do you do? You save your own child. Why? Because love is an extremely powerful emotion and it makes people highly irrational, sometimes at the expense of others' well-being.

One often hears the expression "I will do anything for..." in combination with a declaration of love. This is a wonderful thing... **if** you are the object the love, otherwise, you might be screwed. The news is filled with people who rob, lie, cheat, steal, and even murder out of love for another person, not out of hatred. We all honor and respect those who love their country so much that they will kill for it... unless of course, if the country they love isn't ours. And there is nothing more beautiful than loving God... unless of course, the person's God is the one that rewards people for flying airplanes into buildings or tells U.S. Presidents to invade countries.[77] People do

---

[77] BBC - Press Office - George Bush on Elusive Peace. (n.d.). Retrieved September 22, 2016, from http://www.bbc.co.uk/pressoffice/pressreleases/stories/2005/10_october/06/bush.shtml

horrible things out of hatred, but they also do horrible things out of love.

In some cases, you are more screwed if you *are* the object of the love. John List was a man who loved his family so much that he killed them all so they would go directly to Heaven, even though he believed he might spend an eternity in Hell.[78] Similarly, Andrea Yates drowned her five children because she believed that she was a bad mother and didn't think her children would grow up properly because of her.[79] In fact, there have been 94,146 known cases of filicide, or the murder of one's own child, between the years of 1976 and 2007.[80] A sampling of these cases indicate that with many of them, the perpetrator claimed that he or she murdered the child/children out of love, and the perpetrator was not found to be criminally insane. But perhaps Jesus, who loves us the most, has the most bizarre way of showing it. According to the Christian Bible's exact numbers, God has killed 2,476,633 people. Adding in estimates for his more famous slaughters such as first-born Egyptians and the entire world in a flood (besides Noah's family), the estimate is about 25 million.[81] As an interesting comparison, according to the Bible, Satan has killed 10 people (and that was because of a bet with God). Jesus loves us so much that he allows unbelievable suffering in the world—much of it due to natural causes. He loves us so much that, according to 70% of Christian Americans[82] he allows the majority of humanity (those who don't

---

[78] John Emil List | Murderpedia, the encyclopedia of murderers. (n.d.). Retrieved from http://murderpedia.org/male.L/l/list-john-emil.htm

[79] West, S. G. (2007). An Overview of Filicide. *Psychiatry (Edgmont)*, 4(2), 48–57.

[80] Murder-Suicides Involving Children: A 29-Year Study : The American Journal of Forensic Medicine and Pathology. (n.d.). Retrieved from http://journals.lww.com/amjforensicmedicine/Fulltext/1999/12000/Murder_Suicides_Involving_Children__A_29_Year.2.aspx

[81] Wells, S. (2007, January 13). Dwindling In Unbelief: How many has God killed? (Complete list and estimated total). Retrieved from http://dwindlinginunbelief.blogspot.com/2007/01/how-many-has-god-killed-complete-list.html

[82] Pew Research Center (2015, November 10). Most Americans believe in heaven ... and hell. Retrieved from http://www.pewresearch.org/fact-tank/2015/11/10/most-americans-believe-in-heaven-and-hell/

accept Jesus as the Lord and Savior) to be tortured in Hell for all eternity. With lovers like Jesus, who needs haters?

What if we love everybody? Anyone who has been fortunate (or unfortunate) enough to experience true love whether it be for a partner, friend, parent, or child, knows that there is a significant difference between that kind of love and the "love" one claims to have for strangers. When one talks about love for humanity, "neighbors," or some stranger they just met, they are talking more about a general respect or appreciation. If you think this is a cynical view and you really think you love everyone, ask yourself why you spent $10 watching that last movie instead of spending it on a starving child that you "love".

Unlike many other species, humans need to be taken care of for many years after they are born. For this reason, love has evolved to ensure the survival of our species. Without love, parents would have little to no interest in expending energy and resources in caring for their young. Other forms of love are a result of social bonding necessary for the general survival of species that relies on others. What we find in nature is that the love we have for others is in direct proportion to the benefit and social utility of that love. For example, we generally have a greater love for our partner, caregiver(s), and children than we do for our uncles, cousins, and friends. This, like our desire to overeat and eat lots of sugar, is a result of our evolutionary past. This doesn't make this kind of love "right" (that would be what is called the *naturalistic fallacy*), it just means that universal love for all humanity where every person is equally "loved" doesn't come naturally for us.

If you are waiting for the answer to world peace, I don't have one. I respect all life, but I prefer dogs to mosquitos, humans to squirrels, Americans to Liechtensteinians, and my children to other people's kids. If I had to cause innocent people pain by robbing a store so my family could eat, I probably would, regardless of what happened to me. This is love. Beautiful if you are my family, but problematic for everyone else and potentially devastating for those who got in my way. Maybe non-attachment and indifference is the solution? Or maybe an imperfect world filled with love and suffering is better than a perfect world filled with indifference.

People Are Much More Selfish Than You Think

Another idea that is heavily romanticized is *altruism* or selflessness. An altruistic act is a behavior that benefits one or more other beings at the expense of the being who expressed the behavior. This is one of those coveted ideas in virtually all societies. We like to imagine that kind people who help others, make sacrifices, risk their lives for others, or even give their lives for others have some kind of special heart or "soul." The fact is, our behavior is controlled by our brains and most of what we believe to be selfless acts are far from selfless. This is actually good news—the more we understand this, the more we realize that we are all capable of doing fantastic things for others and we will be motivated to do these things because it will directly improve our own well-being.[83] Selfish desires can result in doing good things for others.

We as human beings are naturally bad at understanding complex behavior, but the human brain is not. The conscious choices we make are difficult to justify at times, yet we feel compelled to take action. Lacking knowledge in the behavioral sciences, we interpret this the best way we know how, or *rationalize* our behavior—usually through simplistic ideas offered by religion or mythology. The "god," "angels," or "kind soul" that we think guides us to give $20 to a homeless person on the street is actually our unconscious brain that knows such behavior will be rewarded with admiration of others, gratitude from those being helped, alleviation of our own guilty conscience, complying with social pressure, and/or perhaps even the belief that we are securing our own place in Heaven. Neurologically, acts of kindness release endorphins into the brain that result in what has become know as the "helper's high." Do not confuse these motivators for reasons— reasons are conscious and deliberate whereas motivators can be unconscious. In other words, one could have nothing but the best and most pure intentions of helping another, but the intentions are a result of the unconscious motivators. Like it or not, we benefit from these "selfless" acts.

---

[83] Weinstein, N., & Ryan, R. M. (2010). When helping helps: Autonomous motivation for prosocial behavior and its influence on well-being for the helper and recipient. *Journal of Personality and Social Psychology, 98*(2), 222–244. http://doi.org/10.1037/a0016984

What about people who literally die for others? How could a brain that is primarily interested in its own survival allow such a thing? In most cases, it doesn't mean to. Most people who commit selfless acts that result in their death are not planning to die; they are counting on surviving. For all the same reasons mentioned above, the benefits outweigh the risks, and the altruistic act is performed. As for understanding the altruistic behaviors of those who know they are going to die, we need to get our science on.

In 1976, evolutionary biologist Richard Dawkins published perhaps the most groundbreaking book in the field since Darwin. In his book, *The Selfish Gene*, he laid out the facts supporting the theory of gene-centered evolution rather than the idea that evolution is centered around the organism or group.[84] This revolutionary concept explained what had previously been a scientific mystery—the ideas of altruism and selflessness explained naturalistically. This idea combined with *kin selection*, *group selection*, *reciprocal altruism*, and *pathological altruism* explains why people might give their lives for others, even if the others are unrelated. An uncomfortable idea within this idea is that **a human is just a gene's way of reproducing itself**. Let that sink in for a moment.

There is a degree of manipulation involved in altruistic behavior. Perhaps the most uncontroversial of this manipulation is the deliberate rewarding of desired behavior and condemnation of selfish behavior through operant conditioning. In business school, people are taught that there are more effective forms of reward than just money such as public recognition. Soldiers are awarded the Purple Heart, Cops are given medals, and kids who rescue cats from trees are featured in their local newspaper. Consciously or not, we all reward selfless behavior because we want to benefit from the behavior when needed. I refer to this as "manipulation" but this may not be entirely fair. When you say "thank you for your service" to someone in our military, the chances are you are not consciously trying to manipulate them into risking their life, so you don't have to risk yours. The odds are, you are simply following a social convention that in itself has become an act of kindness. These social conventions can be considered *evolved group*

---

[84] Dawkins, R. (1990). *The Selfish Gene* (2 edition). Oxford ; New York: Oxford University Press.

behaviors that, consistent with evolutionary theory, have evolved because they benefit the group. For the soldier, being publicly recognized fills a strong psychological need of importance and meaning that serves as a strong motivator for the risky behavior.

The fact that we almost always act out of self-interest does not make us flawed, "fallen," or evil; it makes us perfect—perfectly human. We don't make decisions with our heart, and none of us has some mystical evil soul that keeps us from doing kind things for others. It is because of our selfish behavior and our selfish genes that we still exist as a species. No need to feel guilty or sorry about that fact. Do something for other people knowing that you will also be helping yourself, and because it simply feels really good to help others.

"Microaggressions" Are Less Common and Less Problematic Than People Think

Unless you have been living in a safe space for the past several years, you undoubtedly heard of the term "microaggression." Although coined in 1970 to refer to negative behaviors of non-blacks toward blacks,[85] the term microaggression has been expanded to mean a subtle but offensive comment or action directed at a minority or other nondominant group that is often unintentional or unconsciously reinforces a stereotype.[86] While there is no question that these kinds of behaviors exist, and the behaviors stem from prejudice or deliberate acts of discrimination, they are far less common than many people think.

From a scientific perspective, I have an issue with the modern definition and usage of "microaggression." The term "stereotype" is used and confused with prejudice or discrimination. *Stereotypes*, academically speaking, are not negative or wrong; they are simply a representation of common traits that we view as characteristic of a

---

[85] Sue, D. W. (2010). *Microaggressions in Everyday Life: Race, Gender, and Sexual Orientation*. John Wiley & Sons.

[86] Microaggression | Define Microaggression at Dictionary.com. (n.d.). Retrieved from http://www.dictionary.com/browse/microaggression

social group.[87] This means that they may be accurate, or may not be, and a reinforcement of a stereotype is not always a bad thing. For example, escape convicts are often seen as dangerous. This is a stereotype. Not all escape convicts are dangerous, but the reinforcement of that stereotype has obvious social utility. Regardless, for the sake of entertaining this idea, let's assume that reinforcing *any* stereotype is a sign of a microaggression because more stereotypes are problematic than not.

> *So the doctor draws trees, "What do you see?" the guy says "sex." The doctor draws a car, owl, "Sex, sex, sex." The doctor says to him "You are obsessed with sex," he replies "Well you're the one drawing all the dirty pictures!"* - **Bob Wiley, What About Bob**

There is a psychological phenomenon known as *pareidolia* where the mind perceives a pattern where none exists.[88] The stimuli could be visual as in seeing Jesus in burnt toast, auditory as in hearing Satanic verses when playing the Beatles records backward, textual as in finding "hidden messages" within newspapers or books, or *contextual* as in seeing prejudice and discrimination in speech or behaviors. This is extremely common in humans, and when combined with the confirmation bias or even a hint of persecution complex, "microaggressions" can be seen everywhere.

In terms of microaggressions, intent is not required because prejudice and discrimination can be expressed subconsciously. For example, if a waiter consistently gives preferential treatment to the white customers without realizing it, the act of giving the black customers inferior service can reasonably be classified as a microaggression. But observing and testing for this kind of repeated behavior is something a social scientist might do, but not your average person. Unless we are conducting a study, it's unlikely that we would define "microaggression" in such a way that there is a clear delineation

---

[87] Nelson, T. D. (Ed.). (2009). *Handbook of Prejudice, Stereotyping, and Discrimination* (1 edition). New York: Psychology Press.

[88] Liu, J., Li, J., Feng, L., Li, L., Tian, J., & Lee, K. (2014). Seeing Jesus in toast: Neural and behavioral correlates of face pareidolia. *Cortex, 53*, 60–77. http://doi.org/10.1016/j.cortex.2014.01.013

between behaviors that are microaggressions and are *not* microaggressions. If we are looking for biases that would strengthen our ideological position, such as "racism is prevalent in the restaurant business," we are far more likely to interpret behaviors through this ideological lens. In other words, the number of behaviors that qualify as a microaggression increase significantly.

A behavior does not become a microaggression simply because one is offended by the behavior. The term implies that there is a guilty party who holds prejudicial tendencies, conscious or unconscious. Biases, unclear definitions, ideologies, and pareidolia aside, statistics alone account for much of what is often mistaken for microaggressions. What if the waiter in our last example had no racial bias, but actually showed favoritism to the tables he thought were the nicest? Let's say fifty percent of the time the customers at the nicer table would be white and fifty percent of the time they would be black. Since microaggressions require they be directed at a minority or other nondominant group, we would record several instances of "microaggressions" against the black customers even though there wasn't the slightest evidence of conscious or unconscious prejudice. No instances of microaggressions would be recorded against the white customers because whites are a dominant group, so by definition are excluded. What we end up with is several cases of perceived discrimination where none exist.

Behaviors that stem from prejudice or deliberate acts of discrimination exist and are too common (is even one acceptable?), but screaming "microaggression" where no such behavior exists is scientifically inaccurate and causes those who are less scientifically-minded to dismiss the phenomenon altogether. Exaggerating the seriousness or prevalence of a social problem is not an effective strategy for solving the problem. Microaggressions exist, but they are not as common as they appear to be.

Religious Ideas Are Protected By Motivated Reasoning More Than Any Other Class of Ideas

If you're part of the one in four Americans that identify as having no religion, then you might be baffled by the alarmingly high number of people in the United States who hold beliefs clearly at odds with scientific facts and statistical probabilities. Today, roughly 40% of Americans believe that God created humans in their present form, 57% believe that Satan exists, and about 41% believe that Jesus will return to Earth within the next 40 years.[89] The reason that these and other similar beliefs continue to be so prevalent in the 21st century is because they are protected by motivated reasoning more than any other class of ideas. Ideas that challenge these sacred beliefs are the most uncomfortable; therefore, they are the most avoided, ignored, and dismissed.

For any given idea that challenges a sacredly held religious idea, there are often several reasons why we avoid exposure to them, refuse to entertain them, and refuse to accept them, regardless of the evidence. These ideas are so well-protected that even good critical thinkers and the scientifically-minded protect their sacredly held religious beliefs through a process called *compartmentalization*, where they are not given the same level of unbiased scrutiny as other ideas. The most common reasons for avoiding, ignoring, and dismissing ideas that challenge sacredly held religious beliefs include:

- Fear of the Slippery Slope
- Fear of Questioning Authority
- Refusal To Question Authority
- Fear of Social Response
- Fear of Entertaining Evil, Sick, or Immoral Thoughts
- We Don't Want To Be Seen As "Unpatriotic"
- We Don't Want To Turn Our World Upside Down
- The Sunk-Cost Fallacy
- Fantasy Over Reality
- We Want a Superhero and a Supervillain
- We Want To Be Special

---

[89] Street, 1615 L., NW, Washington, S. 800, & Inquiries, D. 20036 202 419 4300 | M. 202 419 4349 | F. 202 419 4372 | M. (2010, July 14). Jesus Christ's Return to Earth. Retrieved from http://www.pewresearch.org/daily-number/jesus-christs-return-to-earth/

- The Work of Satan

Adam, Eve, and the 6000 Year Old Universe

Let's take the dangerous idea that the universe is actually more than 6000 years old and Adam and Eve is a fable (i.e., a short tale to teach a moral lesson, often with animals or inanimate objects as characters). Works of fiction often have several "plot holes" that make the story more appealing, but only if you don't think about the plot holes. For example, why did God put the tree of knowledge in the Garden of Eden? Did God not know they would eat from it? If Adam and Eve did not know about good and evil before they ate the forbidden fruit, how were they supposed to know that not obeying God was evil? Plot holes aside, a believer of the literal Adam and Eve might reflexively conclude that the ideas that the universe is actually more than 6000 years old and Adam and Eve is a fable is the **work of Satan** and **fear entertaining this evil idea**. Many devoted believers accept the Bible as "God's Word," meaning that questioning the authenticity of the Bible is **questioning the authority of God**, which they might not do out of fear or devotion. If the believer were to find the idea of an old universe where humans were the product of evolution compelling, due to the **sunk-cost fallacy**, he or she might reject the idea reasoning that maintaining the 6000-year-old universe belief is warranted because the resources already invested in the belief will be lost otherwise. In addition, he or she might realize that they are on a **slippery slope**. After all, if there was no Adam and Eve, then there was no original sin, and Jesus died for nothing. Or if the Bible was wrong about Adam and Eve, then what's with the genealogies in the New Testament going back to Adam? If the Adam and Eve story is a fable, why is the story about Jesus coming back from the dead and floating "up" into Heaven to be taken as historic fact? One uncomfortable thought opens an entire box of even more uncomfortable thoughts. If we manage to escape that concern, we still have a host of consequences we would need to deal with if we were to accept this idea based in reality. First, we would need to deal with the **social response** from our church, family, and friends who are likely to all hold the same creationist ideas. In some cases, this can tear a family apart. If accepting the old universe idea did lead to no longer believing in Christianity, the believer might **fear being seen as unpatriotic**. The validity of associated ideas such as eternal life and having a personal relationship with the creator of the universe would be in question, which would be a tough pill to swallow on top of his or her **worldview already being turned upside down**. The reasons, both conscious and unconscious, regularly prove to be too great to overcome, and the uncomfortable idea of a 13.8 billion-year-old universe is avoided, ignored, or dismissed—regardless of the evidence.

Using the reasons previously mentioned in the bullet points, see what reasons might keep you or others from entertaining and accepting the following uncomfortable ideas (uncomfortable to at least 75% of the United States population).

The Soul

There is no such thing as a soul. Given everything we know about the human brain, it is clear that the mind is a function of the brain. We can control thoughts and feelings by manipulating the brain through surgical and chemical intervention. Memories, what we consider a core aspect of the self, are unquestionably stored in the brain. Diseases such as Alzheimer's deteriorate the brain, memories disappear, and the person often becomes unrecognizable to family and friends.

The concept of the soul within religion doesn't make sense. The concept of the soul allows for religions to claim eternal life using the punishment of Hell or reward of Heaven as ultimate motivators. But if we are all born with perfect souls, what is it about a person that makes them a good or bad person? There are environmental and biological factors. Do we blame babies for being born with the inability to empathize? Do we blame babies for being born into poverty to a crack-addicted single mother? At some point in one's life, an "evil" person has to have his first evil thought, take his first evil action, or make his first evil decision. Even if we imagine a magic force called "freewill," what makes that child use his "freewill" for evil—assuming the soul was perfect? If it was the child's environment or biology, can he be blamed? If not, what's left? How is it the "soul's" fault that a person is "evil" or "good"? Of course, we can say that some people are born with evil souls, but that would make God a monster.

When we die, that's the end for us. "We" are an incomprehensibly complex collection of biological components that ultimately give rise to what we call consciousness through a process called *emergence*. In its simplest form, emergence is how we get "wetness" by combining two gasses—hydrogen and oxygen. We can't find wetness in hydrogen, nor can we find it in oxygen. Amazing properties emerge where we see greater complexity, and there is nothing more complex than the human brain (that we know of). Like when an H_2O molecule is split, the emergent property (wetness) ceases to exist—it doesn't "go" anywhere. Likewise, when our biological brain dies, the emergent property we call the "mind" ceases to exist—it doesn't "go" anywhere.

The Christian Bible

There are literally hundreds of holy texts from dozens of religions around the world, most of which are claimed to be divinely written or inspired.[90] All the religions have competing and incompatible ideas. Lacking education in world religions, many people are not even aware of these other texts and are convinced that the holy book of their particular religion, from their particular culture, is the one "true" holy book.

The Bible is like a textual Rorschach test where ambiguous text can be interpreted virtually any way. This explains why there are thousands of Christian denominations and people's concept of God perfectly reflects their moral, political, and philosophical views. Should we be killing gays like God commands in the Old Testament? It depends on how you feel about gays. If you like them, you can say that Jesus is "the new covenant" and the old rules don't apply. But then what about the Ten Commandments? For virtually every verse or story that can be used to justify a position, another can be used to justify the opposite. This is not generally disputed since the answer to someone who has a different biblical justification than you is "you're taking that out of context." This response works because "in context" often means you have to understand the entire Bible and the will of God the same way as the person claiming that you are "taking that out of context."

---

[90] Religious text. (2016, September 24). In *Wikipedia, the free encyclopedia*. Retrieved from https://en.wikipedia.org/w/index.php?title=Religious_text&oldid=740908920

Unsure about this? Google "is god against homosexuality" and skim through the top results.

The "Goodness" of the Biblical God

"The God of the Old Testament is arguably the most unpleasant character in all fiction: jealous and proud of it; a petty, unjust, unforgiving control-freak; a vindictive, bloodthirsty ethnic cleanser; a misogynistic, homophobic, racist, infanticidal, genocidal, filicidal, pestilential, megalomaniacal, sadomasochistic, capriciously malevolent bully." **– Richard Dawkins**

It is shocking to consider that only about one in five Americans have read the Bible from cover to cover.[91] Church leaders are very selective about which passages are read in church—usually the "good" verses from the New Testament. But there are 31,102 verses in the Protestant Bible. This is like a used car salesmen telling you all about how great the radio is while overlooking that the rest of the car is a defective piece of trash. Here are a few verses you probably didn't hear in read church. Think about these verses and think if we should accept this book as the perfect objective standard of morality.

Slaves, obey your earthly masters with respect and fear, and with sincerity of heart, just as you would obey Christ. (Ephesians 6:5)

When a man sells his daughter as a slave, she will not be freed at the end of six years as the men are. If she does not please the man who bought her, he may allow her to be bought back again. (Exodus 21: 7-8)

Slaves, submit yourselves to your masters with all respect, not only to those who are good and considerate, but also to those who are harsh. (1 Peter 2:18)

A woman should learn in quietness and full submission. I do not permit a woman to teach or to have authority over a man; she must be silent. (1 Timothy 2:11-12)

---

[91] The Books Americans Are Reading. (n.d.). Retrieved from https://www.barna.com/research/the-books-americans-are-reading/

Wives, submit to your husbands as to the Lord. (Ephesians 5:22)

If anyone comes to me and does not hate his father and mother, his wife and children, his brothers and sisters—yes, even his own life—he cannot be my disciple. (Luke 14:26)

Whoever sacrifices to any god other than the LORD must be destroyed. (Exodus 22:20)

This is what the Lord Almighty says ... 'Now go and strike Amalek and devote to destruction all that they have. Do not spare them, but kill both man and woman, child and infant, ox and sheep, camel and donkey.' (1 Samuel 15:3)

Happy is he who repays you for what you have done to us / He who seizes your infants and dashes them against the rocks. (Psalm 137)

If a man beats his male or female slave with a rod and the slave dies as a direct result, he must be punished, but he is not to be punished if the slave gets up after a day or two, since the slave is his property. (Exodus 21:20-21)[92]

I encourage all Christians to look up how other Christians attempt to explain these verses in order to maintain the comfortable idea that God is perfectly good, or that everything in the Bible was written by God and not just people claiming divine authority as a way to enforce their laws. Now that you know how our brains protect us from uncomfortable ideas, you will be able to see this in action in the explanations. This is one example taken from an apologetic (defender of "the faith") website that attempts to explain why that last passage about beating your slaves is consistent with a good god:

The Bible NEVER condones beating a slave, hitting a slave, and never suggests to treat them in a cruel way. The verse in question is dealing with the penalty of such activity, not condoning it or making moral statements about it. Much like how we have laws against rape, domestic violence, and the

---

[92] For over a thousand more Biblical verses that will make any decent person uncomfortable, visit EvilBible.com.

> *penalties for such crimes. The chapter itself implies that the activity is wrong within its own context.*[93]

By focusing on what a verse does not explicitly say, one can redirect your attention away from what the verse does explicitly say and clearly imply: **that beating your slave as long as he or she does not die within a day or two does not warrant punishment.** If this law was anything like the domestic violence laws we have today as the apologist suggests, our domestic violence law would read something like this:

> *When a husband strikes his wife with a rod and the wife dies under his hand, he shall be arrested. But if the wife survives a day or two, the husband is not to be arrested, for the wife is his property.*

The entire foundation of Christianity is absurd. Sometimes satire is more effective than a rational argument because satire lowers many of our defenses that cause us to dismiss uncomfortable ideas. Here is one definition of Christianity that highlights this absurdity (author unknown):

> **Christianity:** *The popular belief that a celestial Jewish zombie who is also his own father, born from a virgin mother, died for three days so he could ascend to heaven on a cloud and then make you live forever only if you eat his flesh, drink his blood, and telepathically tell him you accept him as your lord & master, so he can remove an evil force from your spiritual being that he put there because an immoral woman who didn't know good from evil, made from a man's rib, was hoodwinked by a talking reptile possessed by a malicious angel to eat from a magical tree.*
>
> *Makes perfect sense.*

---

[93] Bible Says It's Okay to Beat Your Slave, As Long As They Don't Die? Exodus 21:20-21? (2013, June 9). Retrieved October 15, 2016, from http://www.revelation.co/2013/06/09/bible-says-its-okay-to-beat-your-slave-as-long-as-they-dont-die-exodus-2120-21/

Belief and Faith

If you were born in the Middle East, to Muslim parents, in a Muslim culture, the chances are overwhelming that you would be a Muslim and die a Muslim. Do you think it is just for a god to judge people based on their beliefs given that belief is strongly correlated with geography and the beliefs of one's parents? Do people deserve to be tortured in Hell for eternity for this reason?

Faith is not a virtue. At least not when others have faith in different things than we have faith in. If we live in ways that are consistent with our "faith," can we blame other cultures for living consistently with their faith? If our faith tells us that Jesus is Lord and gays should be stoned to death because the Bible says so, can we blame a Muslim for claiming Allah is Lord and infidels should be killed because the Koran says so? Once we try to legitimize our faith with "evidence" or reasons, we not talking about faith anymore; we are talking about justified belief (justified from the perspective of the one justifying). If we have reasons for believing an idea, we don't need faith. But Christianity requires faith.

Some argue that they have justified reasons for their belief, and through faith that belief is strengthened. So rather than using faith in place of reasoning in the acceptance of an idea, faith is used in place of reasoning in the strength of belief in the idea. While this may fill the Christianity requirement loophole, it still makes uncomfortable ideas that are antithetical to religious belief immune to reason and evidence. This may not be a problem for you if you are a Christian presented with an atheistic idea, but surely you can see that this is a problem when attempting to reason with Islamic extremists who, through faith, strongly believe the idea that their god wants us dead. Your faith may be harmless, but your support of the concept of faith as a legitimate way to know things, is not.

Being an Atheist Doesn't Make You Smarter and Certainly Not Better at Critical Thinking

There are literally dozens of studies and scientific polls that look at intelligence and religious belief. When evaluating all the available research, we find that there is no compelling evidence that atheists are more intelligent than theists. Even if one were to accept some of the studies that showed that atheists were overall more intelligent, these are statistical conclusions, meaning that you can easily be one of the moronic atheists.

Atheism is simply the lack of belief in any gods. One of the suggested reasons why some claim atheists are more intelligent is because atheists critically examined their beliefs at some point and can intelligently support their belief in a naturalistic universe. There are two problems with this reasoning: First, many atheists are atheists because they were raised by atheist parents—they never examined their beliefs. Second, many theists also critically examined their beliefs at some point. Identifying those atheists who aren't very good critical thinkers can usually be done by looking at their arguments. Sticking with uncomfortable ideas, here are some bad atheist arguments debunked presented in idea form that will be uncomfortable to many atheists.

There is Evidence for God

People who don't know that there is a difference between "proof" and "evidence" cannot be good at critical thinking. Knowing what constitutes proof versus evidence is more difficult, and is context-dependent. As a general rule of thumb, we can prove something mathematically or deductively through logic. In legal terms, "proof" is really just strong evidence. What is *evidence*? Anything presented in support of an assertion. Evidence could be either strong or weak. The apparent design of natural objects could be presented in support of a creator god, i.e., evidence for God. One can argue that this is weak evidence, but it is still evidence. The more evidence that is presented, the stronger the overall evidence is. A critical thinker needs to acknowledge this and compare the evidence for a god to the evidence against a god.

No, Believing in God is Not the Same as Believing in Santa Clause.

I know, Santa and God have many similarities. They both perform miracles, are ageless, are givers of gifts, have white hair and beards, see you when you're sleeping, know when you're awake, know when you've been bad or good... but there is one big difference: Santa is *falsifiable*. We can know that Santa doesn't live in the North Pole. We can know that Santa does not break into all Christian kids' homes and leave gifts. Without doing these things, Santa is no longer Santa, therefore, cannot exist. Even if you can muster some philosophical argument that one can't know Santa's not living underground somewhere in the North Pole or that maybe Santa only delivers to a few kids, there is still the fact that about 86% of the world's population believe in some form of god or gods and virtually no adults of normal mental ability believe in Santa, which means there is obviously a major difference and believing in Santa is not "the same" as believing in God.

Your Examples of History's Jesus-like Figures are Likely Made Up or Greatly Exaggerated

This can be a really strong argument but is rarely presented as such. Atheists generally quote exaggerations of these stories from atheist websites or reference literature written after the story of Jesus. They can rarely point to any authentic documents that predate the New Testament. Basically, they are making up Jesus-like figures. The strong, yet rarely used version of this argument points to the early Christian apologists who understood that the Jesus story looks to be just another of the many similar stories told at the time and explained it as the work of the devil.

Early Christian apologist, Justin Martyr (100-165 CE), used these similarities in his First Apology (Chapter 22 — dated about 150-155 CE) to help win pagan converts who already accepted the idea of virgin births, crucifixion of man-gods, and miracles. Another Christian Father, Tertullian (160-220 CE) wrote in Chapter 21 of his Apology about the similarities with pagan god Romulus. It is clear that in these early writings both Christian Fathers accepted, and sometimes even promoted, the similarities to Jesus with other gods and man-gods. So how do they and other early Christian apologists explain these extraordinary similarities? Diabolical Mimicry (also known as demonic imitation). This is the idea that Satan (the devil) started all these rumors about other gods long ago in anticipation of Christ—just so people would have a reason to question the authenticity of the Jesus story (as explained by Justin in Chapter 54 of his apology)[94].

Evolution Does Not Answer the Question of Where we Came From

Evolution addresses how we evolved, but does not address the origin of life.

It is Foolish To Demand That Believers Prove That God Exists

Recall the difference between proof and evidence. Justified belief requires evidence, not proof. Also, one's lack of ability to prove that something exists says nothing about the thing actually existing.

No, Theists Will Not Understand Why You Don't Believe In God When They Realize Why They Call Zeus a Myth

Similar to the Santa defense, Zeus is a very specific god that is far more difficult to believe in our modern world than the God of theism since Zeus is said to have a physical presence and live on Mount Olympus. This is an especially weak argument when addressing believers in a generic God with few, non-falsifiable properties.

Everyone is Not Born an Atheist

Science clearly shows that we have many natural or instinctual biases that lead to belief in the supernatural. Religion appears to be a natural conclusion to us. We are likely to detect agency where none exists,[95] religion comes naturally to young children,[96] dualism emerges as an evolutionary accident,[97] and we have literally dozens of

---

[94] Bennett, B. (2010). *The Concept.* eBookIt.com.

[95] NeuroLogica Blog » Hyperactive Agency Detection. (n.d.). Retrieved from http://theness.com/neurologicablog/index.php/hyperactive-agency-detection/

[96] A reason to believe. (n.d.). Retrieved from http://www.apa.org/monitor/2010/12/believe.aspx

[97] Schloss, J., & Murray, M. (2009). *The Believing Primate.* Oxford University Press. Retrieved from http://www.oxfordscholarship.com/view/10.1093/acprof:oso/9780199557028.001.0001/acprof-9780199557028

cognitive biases that lead to religious belief. It is much more accurate to say that we are all born believers—it is the specific details of religions that were not born with. Those have to be taught.

Most Apparent Bible Contradictions Can Easily Be Explained

So many atheists look foolish attempting to trap Christians with Bible contradictions, especially when those Christians never claimed that the Bible was a flawless document. Even when certain Christians do claim that the Bible is flawless (i.e., they believe that the Bible is *inerrant*), any knowledgeable Christian will be able to explain away apparent contradictions. Many of the proposed contradictions are not contradictions and do have good explanations, which just makes the atheist look dumb. There are proposed contradictions where the explanation appears far less likely than accepting that the Bible isn't a flawless book but rather a collection of 66 books written over hundreds of years by many anonymous authors with different beliefs and understandings, but unless you know what those are, avoid this argument.

In addition to bad atheist arguments, atheists can also hold positive beliefs about conspiracy theories, alien abductions, ghosts, ESP, psychic healing, talking to the dead, witches, magic, reincarnation, and other forms of woo that are not related to a god or gods. Atheists can be anti-science, as well. The lack of a belief in any gods says very little if anything about intelligence, and atheists are not immune to stupidity or gullibility.

You Should Give President Trump or President Clinton Your Support

Note: This idea was written on October 1, 2016, before the election at a time when both candidates were very close in the polls.

If you are truly one of the rare moderates who think either candidate will make a good President, then you are probably lying. If not, then this idea shouldn't be uncomfortable for you. Statistically, however, you are most likely to have very strong feelings for both of the candidates; but only positive feelings for one of them.

As I disclosed in the preface, I am left of center politically. In November, I plan on voting for Clinton and can't imagine voting for Trump. Yet if he wins, he will have my support. Hillary Clinton is my candidate, but if Trump wins, he will be my President.

So if Trump wins, how will I give him my support? Will I support his proposed policy of building a wall to keep Mexicans out? Will I support keeping Muslims out of this country? Will I support the killing of terrorists' families? No, no, and hell no. I will not support every decision he makes nor every policy he proposes, but I will do him and our democracy a far greater service. I will support his Presidency and the integrity of the office by giving him a fair shot to prove me and what would have to be the minority of electoral delegates wrong about his ability to be a good President.

I spent the last years, as did many of you, watching the citizens of this country, special interest groups, and the conservative media demonize and even dehumanize President Obama. I have seen so much hatred from people close to me it has literally made me sick on several occasions. I'm not referring to the justified responses to actual wrongdoings or legitimate criticisms well argued by those who were opposed to Obama's actions and policies. I am referring to the plethora of ignorant exaggerations fueled by bitterness and animosity. I'm referring to the unfair attacks on the person that we, through the democratic process that we love so much that we force on others, have elected to lead this country. Politics aside, "hating Obama" provided fodder for shock jocks and media who preyed on the anger, hate, and fear of the American public. It's all part of a pathetic game where we, the American public, are always the losers by being an embarrassment to the rest of the world.

Here is how you, as an American citizen, should support our President, no matter who they are or how awful you may think he or she is.

- **View the President as part of your ingroup.** No longer is he or she the other party's candidate, but he or she is now your President. We are all in this together. Making this ingroup salient will go a long way in removing many of the prejudicial biases that we might have toward our President.

- **Consistently question your confirmation bias.** If you believe the President is evil, you will tend to ignore all the facts that contradict your belief. You will tend to remember only those facts that support your belief, and forget those that don't. You will interpret the actions of the President in such a way that confirms your bias. In other words, your impression of the President will never be fair or accurate unless you realize how the confirmation bias is affecting your judgment.
- **Get your news from unbiased sources, or at least left- and right-leaning sources.** If you can't do without FOX News or the Huffington Post, then check them both for news and check the facts for yourself. Otherwise, get your news from the Associated Press or NPR... and still check the stories for facts and spin. Don't allow yourself to be manipulated. Think critically!

Supporting your President does not mean blanket agreement with his or her policies and vision for America. Speak up and speak out when you disagree; this is both your right and your duty. But before you become active in the political process, even if just social media warrior making your opinion known to your family and friends, remove the plank from your own eye (thank ya Jesus) by offering unbiased, fair, and helpful criticism based on facts and critical thought. This is how you can support your President and your country.

If You're Offended, You're Part of the Problem

Here is a very important fact that very few people seem to understand: offensiveness is not a property of an object, whether it be a word, comment, photograph, gesture, or idea. Nothing is inherently offensive; **it is our interpretation of the stimuli that makes it offensive to us.** Eleanor Roosevelt once said, "No one can make you feel inferior without your consent." The same holds true for offense. No one can offend you without your consent.

Offense is not a reaction; it is a response that requires deliberate cognitive processing at which point you decide to be offended or not. It is a choice, categorically different from physiological reflexes and primary emotions such as fear or anger that occur without thought. In looking at the phenomenon of offense, we can look at the conscious reasons why we choose to be offended and look at the unconscious reasons behind the choice.

Why We Choose To Be Offended

Nobody thinks they are wrong for being offended. If you're offended, in your mind, your offense is an appropriate response to the stimuli (e.g., comments, gestures, jokes, ideas, photos, words, books, etc.) that is deliberately or unknowingly causing anger, annoyance, or uncomfortableness (i.e. harm) to you or someone else. If someone were to ask you why you were offended, you would be able to give a reason, which would likely be one the following.

- **Malicious intent is suspected.** If we think someone is deliberately trying to cause us or someone else harm, we take offense. Malicious intent is often found in deliberate ethnic slurs said with anger or in passive-aggressive attempts at humor.

- **Ignorance is suspected.** People can trigger offense without knowing it—like granddad who innocently refers to an entire ethnicity as "those people." We wouldn't necessarily blame the person, but we do want to make it clear that the ignorance being unknowingly spread is hurtful to others.

- **Implicit bias suspected.** Implicit is another word for unconscious, or not done knowingly. It is the fear that prejudice, discrimination, or a negative stereotype is being promoted. An example might be a man holding a door open for a woman where the woman is offended because she suspects the man's actions are a result of a male-dominated society in which women are inferior and in need of help.

Why do we experience the anger, annoyance, or uncomfortableness that triggers the offense? Why might one person of Polish heritage laugh at a Polish joke, and another person of Polish heritage find the joke offensive?

The Unconscious Factors That Influence Our Decision To Be Offended

All of us are unique. Every person on the planet has differences in their biology as well as different life experiences. As one example, people are all over the scale in terms of emotional sensitivity, which plays an important role in taking offense and has both genetic and environmental components.[98] If we understand this, we are less likely to have negative feelings towards people who we feel are too easily offended, or people who we feel too often offend others. But there are many reasons why we get offended that are unfair to others, and we make the world more hostile place rather than a better one. We can do something about this by understanding these unconscious factors and using our metacognition (i.e., thinking about how we think) the next time we are deciding to become offended or not.

Here are some of the unconscious factors that influence our decision to be offended.

- **Confirmation bias.** If we are convinced that a stimulus is offensive, we will interpret the stimulus in such a way where we are more likely to find it offensive. For example, if we believe we live in a world where men frequently show their dominance over women, then when we see a man holding a door open for a woman, we will see that as a display of dominance because the "weak and fragile" woman can't open her own door. At the same time, we will ignore the fact that the same man also held the door open for the man who walked through the door before the woman.

- **Self-victimization.** The process of attributing your failures to someone else when such attributions are unjustified is known as *self-victimization*. Self-victimization is commonly used to reduce cognitive dissonance. For example, if we think we are a smart and hardworking person yet we are struggling professionally, we might point to a work environment that is "rigged" against us when no evidence supports such rigging. The self-victimization

---

[98] Sensitive? Emotional? Empathetic? It Could be in Your Genes - Stony Brook University Newsroom. (n.d.). Retrieved October 6, 2016, from http://sb.cc.stonybrook.edu/news/medical/140623empatheticAron.php

defense is also used for attention and sympathy. "That dead dog joke offends me! I once had a dog that died. How can you be so insensitive!"

- **Habit.** We can easily slip into the habit of looking for offense in just about everything, especially when our public outrage is rewarded with support from others who are offended. And because of the confirmation bias, we will find something to be offended by.
- **Low self-esteem.** Low self-esteem and paranoia lead to a hypersensitivity when it comes to evaluating stimuli for offensiveness. If we think we're fat, and we think others marginalize us because we're fat, then we are more likely to see malicious intent or bias in stimuli related to weight.
- **Hatred and anger.** Hatred and anger, like other strong emotions, cloud our judgment and contribute to demonizing stimuli. If we hate Republicans, then we are more likely to interpret their message as offensive even if the exact message comes from a Democrat.
- **Moral superiority.** When we feel we have the moral high ground, we like to show it. Calling someone out on being "offensive" is another way of saying "I am morally superior to you—at least on this issue." We want to feel this sense of superiority even if we have to unnecessarily embarrass or call out the other person for their "moral deficiency."
- **Sense of injustice.** "Justice" is another very subjective idea where one person's justice is another's injustice. If a stimulus challenges our sense of justice, we are offended. We might have good reasons to sense injustice, or we might simply fail to realize that what we perceive as injustice is really just a difference of opinion based on prioritizing values differently.
- **Hero complex.** Everyone is the hero in their own life narrative. We all want to be the hero that saves the poor, marginalized group from destruction. Even the Ku Klux Klan gets offended. A KKK leader expressed his outrage when a Virginia mom dressed her 7-year-old in a Klan costume for Halloween. The Klan leader

said that the costume was an important symbol and not to be taken lightly.[99]

The Person/Idea Distinction Myth

Some people think that they are off the hook for being offensive if they criticize the idea and not the person. This is not the case. We form our identities or self-concepts largely using the ideas in which we believe. By attacking the pro-life position, you are attacking the person whose morals, standards, and principles are inextricably linked to the pro-life position. If you ask that person to describe themselves, they might start with "I am a pro-life advocate." This certainly doesn't mean that criticizing ideas is wrong; it just means you can't blame people for taking offense solely because your opinions are directed at the idea, not the person.

The Optimal Strategy

Some people hold the idea that we should adjust our behavior to satisfy the most easily offended people, some think we should not worry about offending anyone in favor of free speech, and some people believe that a balance is an optimal strategy. Some people are okay with offending others outside specific social groups but not within others. For example, an atheist may be fine with ridiculing Christian beliefs and at the same time being offended by those to ridicule Muslim beliefs. Some people have a higher tolerance for offense in certain situations such as being fine with a comedian saying prejudicial things at a comedy club but not the same things being said at a PTA meeting. And some people can get away with things that would be deemed "offensive" if someone else were to say or write it due to the fact that we hold certain people to higher standards. For example, nobody cares much what some obscure blogger writes compared to a Presidential candidate.

People are going to get offended. They have a right to get offended just as those who offend have a right not to care. Because most of us have values, principles, and standards that are not objectively better or

---

[99] Abad-Santos, A. (2013, November 6). KKK Imperial Wizard Hates to See Kids in Klan Costumes: "There"s No Respect'. *The Atlantic*. Retrieved from http://www.theatlantic.com/national/archive/2013/11/kkk-imperial-wizard-says-kid-klan-costume-disrespectful/354815/

worse than other people's values, principles, and standards we are going to have different thresholds for what reasonably constitutes offense. If you marginalize people who are different from you, expect to feel the wrath. If you communicate with other people in any way, expect to feel the wrath, as well. This is life.

It is Okay to Change Your Mind

There is a story attributed to several people in the last century about changing one's mind. Regardless of who first said it, the point made is a strong one. The story involves an accusation of inconsistency and goes something like this:

Interviewer: *You have been quoted previously as being in support for this position, now you no longer support it. Don't you think you're being inconsistent?*

Interviewee: *When the information changes, I update my conclusions based on the new information. What do you do?*

It seems as if berating politicians and leaders is a favorite pastime here in America. Those who are anti-Obama like to remind us of the time when Obama was against gay marriage. During the 2016 Presidential election, videos lampooning Trump's different opinions over time were not in short supply. In the 2004 Presidential election, John Kerry was synonymous with the term "flip-flopper." Those who are unfamiliar with the scientific method commonly criticize science for "changing its mind," referring to the seemingly never-ending stream of new studies that appear to contradict the scientific norm. This obsession with consistency is part of our personal lives, as well.

In his book, *Influence: The Psychology of Persuasion*, Robert Cialdini discusses the research that demonstrates our desire for commitment and consistency. He writes, "Once we have made a choice or taken a stand, we will encounter personal and interpersonal pressures to behave consistently with that commitment. Those pressures will cause us to respond in ways that justify our earlier decision."[100] Not only do we desire consistency from ourselves, but

---

[100] Cialdini, R. B. (2006). Influence: The Psychology of Persuasion, Revised Edition (Revised edition). New York: Harper Business.

this desire extends to the behavior we expect from others. The problem is, this desire is just another of our heuristics in place to conserve mental energy at the expense of making calculated, rational decisions.

It's not unreasonable to be initially suspect of someone who claims to have changed their mind. Not everybody who claims to have changed their mind has updated conclusions based on new information. Perhaps they are

- **pandering to different audiences** - This is a common tactic in politics where a person will change their views depending on their audience. Kids might tell their friends that they like to drink alcohol while at the same time telling their parents that they don't like alcohol. While pandering could be outright lying, milder forms could include telling half-truths or adjusting the level of commitment one has to an idea. In any case, pandering is a form of deceit and should be interpreted as a negative.

- **confusing noise with the signal** - Those who don't have strong critical thinking skills have a difficult time judging the quality of information that may or may not warrant an updated conclusion. For example, you might accept that climate change is a serious problem that needs to be addressed, but you come across a blog post on bucks-government-coverups.com that explains how climate change is really just a hoax by the Chinese. Unable or unwilling to fact check, you change your mind and decide climate change is a hoax. This "flip-flopping" might frequently happen when you come across a piece of information that shouldn't justify a complete change of view, but does anyway.

- **bad at making good initial choices** - People who frequently change their minds may be doing so because their initial view was based on poor information, and rather than suspending judgment until more information is available, they hastily form an opinion when such an opinion is unnecessary. Once information does become available, they correctly evaluate the information and form a reasonable opinion. In such cases, it is not the update of the opinion that is problematic; it is the commitment to the initial opinion that is the problem.

The key point is that the negatives associated with changing one's mind all depend on the reasons. Reactively, we can't know the reasons since knowing the reasons take deliberate cognitive thought.

We need to accept the uncomfortable idea that we are not right all the time, and that admitting we were wrong brings us one step closer to actually being right. Those who "stick to their guns" in the face of overwhelming evidence and relevant information as it becomes available might be doing so as a way to protect their self-esteem or create a public image of infallibility. We should want leaders who update their conclusions based on new information. We should want scientists to tell us when new evidence is found even it is evidence against an idea that we already accept as true. And we should want our friends and family to make choices and live their lives based on the best information available.

We are our minds. When our minds change, we change. Every idea, opinion, belief and value we have that changes, changes who we are even if in the smallest of ways. We become better people by making an effort to understand the world in such a way that promotes universal well-being. This involves changing our minds at times and accepting uncomfortable ideas.

Concluding Thoughts

You have been exposed to hundreds of uncomfortable ideas in this book, and hopefully, you have entertained rather than dismissed them. But most importantly, you learned why uncomfortable ideas are so difficult to entertain and accept, and this should have given you a greater appreciation for other views and opinions and those who hold them.

It has been said that everyone is entitled to their own opinions but not their own facts. Some ideas are simply factually incorrect, but we won't know this unless we entertain the idea first. However, the vast majority ideas being discussed are those with no obvious right or wrong position. With ideas such as these, it is expected that people will hold extreme views and as a reaction, others will hold views on the opposite extreme. The ideas presented by both sides are often uncomfortable to the other side. This doesn't mean that the "correct" views are in the middle somewhere; it just means that in the middle somewhere is where compromise is often found. When we are dealing with ideas that have no clear right or wrong position, we need the uncomfortable ideas presented by people all over the spectrum to keep our ingroup biases from spiraling out of control and causing us to move towards more unreasonable positions. In a sense, this is a naturally occurring checks and balances system found where uncomfortable ideas are welcomed and entertained.

We are not purely rational creatures. Without emotion, we would not be able to function or continue to survive as a species. We correctly make many life decisions mostly from emotion such as choosing our life parter, what sports and activities we do, our entertainment, and most importantly, having children. But having access to more information and accurate information in these areas will only help us to make better decisions. The uncomfortable ideas we entertain will provide us with that access.

Should we purposely avoid certain uncomfortable ideas for the good of humanity or for the good of our personal well-being? Maybe. Human behavior becomes increasingly difficult to predict the further we look in the future and the more variables we add in the equation. What would happen if we encouraged people of faith to entertain ideas that might convince them their god probably does not exist, and lacking a naturalistic understanding of the world experienced a drastic reduction in their personal well-being? What would happen if we convinced Americans that they are being manipulated to support and even participate in war, and as a result we no longer had a strong military? What would happen if people agreed that morality was functionally democratic yet didn't understand the complex basis of well-being, and used that as an excuse to murder, rape, and steal? Without knowing how to think, maybe some people really are better off just being told what to think. I just hope you're not one of those people.

Part V: Uncomfortable Questions

To get the most out of the questions in this section, ask them to a group of religiously, politically, racially, and sexually diverse people. These make great posts on social media, discussion topics within the college classroom, or questions to ask friends at a party... as long as you don't mind heated discussions. Most of the questions have an implied "why or why not?" Each question is meant to be a discussion, not just a one word answer.

Life Partners

1. Is there a perfect person for everyone?
2. For those who are compatible with others, is there more than one "special" person?
3. Is there anything morally, ethically, or functionally wrong with polyamory (a relationship with multiple, consenting, adult partners)?
4. Aside from being illegal, is there anything wrong with polygamy (having multiple wives) or polyandry (having multiple husbands)? Assume all are consenting adults involved who are aware of the arrangement (i.e., no deception or cheating). Also, assume this act takes place in a culture where neither the women nor men are forced or pressured into the relationship.
5. Can one truly love their spouse and still cheat on them? "Cheating" is defined as a husband or wife having sexual intercourse with someone other than their spouse without the knowledge and permission of their spouse.
6. Can one love more than one person romantically at the same time?
7. Can a marriage or long-term relationship end in divorce/break up and still be successful?
8. Is divorce itself morally wrong?
9. Is it better for children to have a father and a mother than two mothers or two fathers (as in same-sex couples)?

Love and Sex

10. Is being gay morally wrong?
11. Is it wrong to be sexually attracted to a post-pubescent child?
12. Are any acts of sex between two consenting adults morally wrong?
13. Should prostitution be illegal?
14. Is prostitution morally wrong?
15. Is telling strangers or acquaintances that you love them more of a sanctimonious act than a sincere expression of love?
16. Can one feel the love of another person and what exactly does that mean?
17. Is love really a force for good?
18. Is it wrong for teens to have sex?

Humanity

19. Is human population control a good idea?
20. Would the world be a better place if all those who were suffering disappeared or never existed?
21. Can one really do anything out of pure altruism, or does every altruistic act have some self-serving element?
22. Can a society survive with only all ambitious people, or does it need unambitious, unmotivated, and uneducated people to take the crap jobs that the others won't do?
23. If someone is suffering and wants to die, is letting them take their own life a good option?
24. At what point is receiving medical assistance "interfering with God's plan" or "interfering with nature"?
25. Can everything be reduced to a rational answer, or is emotion needed?

26. In the future, if every part of you could be replaced with cybernetic parts, and you were replaced part by part, would you still be you?
27. Does it matter to you what happens to your body after you die?

'Murica

28. Is the Constitution outdated and in need of an overhaul?
29. Would we want the Founding Fathers running our country today?
30. Do more people join the military because they can't get a better job or do more people join because they are heroes and patriots?
31. Should we be teaching our children to pledge their allegiance to our country?
32. Should those who are religious be asking God to bless humanity, not just America?
33. Is an American life worth more than an African life?
34. Is America the greatest country in the world? If so, based on what metrics?
35. Do you think the military demonizes the enemy (i.e., tells lies about them, exaggerates their faults, ignores the good they do) so American soldiers can be more motivated to kill?
36. How was American better in the past than it is today?
37. How is American better today than it was in the past?
38. If you could bring back an American legal and political system of any year, when would it be and why?
39. Is it acceptable to require others to follow your moral, religious, or philosophical beliefs and if so, when?
40. Do all Americans have an equal opportunity to live "the American Dream"?
41. Are some people more disadvantaged than others?
42. Should the government help those who are currently disadvantaged, and to what degree?
43. Should all recreational drugs be made legal?

44. Is democracy the best form of government?
45. Are dictatorships and monarchs superior to democracies providing the leaders are kind and good leaders?
46. If you are politically ignorant, should you vote?
47. What is the difference between democracy and decision by majority? What happens if morons make up the majority of our country?

Faith, God, and Religion

48. Is there such thing as a soul?
49. Do other species have souls?
50. If we are all born with perfect souls, what is it about a person that makes them a bad person, and should they deserve to go to Hell because of that?
51. If some people are born with evil souls, is that their fault or God's fault?
52. If you were born in the Middle East, to Muslim parents, in a Muslim culture, do you think you would be anything but Muslim?
53. Do you think it is just for a god to judge people based on their beliefs given that belief is strongly correlated with geography and the beliefs of one's parents?
54. Why did God put the tree of knowledge in the Garden of Eden?
55. How can God blame us for our imperfections when he made us that way? (perfect people don't make horrible decisions that curse mankind)
56. If we admire Christians who live in ways that are consistent with their "faith," can we blame other cultures for living consistently with their non-Christian faith?
57. Why is faith a good thing to have?
58. How do we know what to put our faith in? Through faith?

59. If one has enough evidence and reason to believe in the Christian God, then what is faith needed for?
60. How much tolerance should we have for a religion we don't believe?
61. Should we live our lives according to the Bible?
62. Is it right for certain religions to encourage large families?
63. Is using birth control morally wrong?
64. Should we be held responsible for the "sins" of our ancestors?
65. Why does human sacrifice of the innocent pay for the sins of the guilty?
66. By what method do we determine which parts of the Bible we should ignore (e.g., kill gays and kids that mouth off to adults) and which we should follow?
67. In the Bible, how can we tell the difference between literal truth, figurative speech, "spiritual" truths, or just beliefs of the authors?
68. If there was no literal Adam, Eve, and talking serpent, what the point of Jesus and why did Jesus seem to think there were a literal Adam and Eve?
69. If Adam and Eve did not know about good and evil before they ate the forbidden fruit, how were they supposed to know that not obeying God was evil?
70. If you didn't believe in any eternal reward or punishment, would you live your life differently and if so how? How do you think your answer affects your morality?
71. What if when we die, that's the end for us? How would your views on this life change?
72. What makes a god a god?
73. Why won't God heal amputees?
74. Why would Jesus want you to eat his body and drink his blood?
75. Why do Christians get divorced, get sick, and die at the same rate as non-Christians?

76. Why did an all-loving God create evil? ("I form the light, and create darkness: I make peace, and create evil: I the LORD do all these things." Isaah 45:7)
77. Why do you think God make the eating of shellfish and ham, or a woman wearing a man's clothing, "abominations"?
78. An all-knowing God knows who will ultimately reject him, so why does a perfectly good God create people who he knows will end up in Hell?
79. Can a mass murderer go to heaven for accepting your religion, while a doctor who spends most of her life healing people in third-world countries go to Hell for not?
80. Is it morally right to have children if you believe that there is a good chance they will end up in Hell?

Metaphysics

81. Is there such thing as "freewill" and if so, what does it mean exactly?
82. What if we are the creations of a more advanced civilization and we are living in a simulated world? What would you do differently?
83. What's the point of life? When you answer, keep asking "so what?"
84. What is the nature of reality?
85. What is truth, exactly?
86. How do we decide what's true if there is no God?
87. How can we tell if something exists or not?
88. Where does logic come from?
89. Where does reason come from?
90. Can something come from nothing?
91. What do you think consciousness is?
92. Try your best to explain the attraction of forces.
93. Is beauty objective? If so, what is the standard of beauty?

94. How does the mind interact with the body?
95. Do you think computers will ever become conscious?
96. What, if anything, can you be certain of?

Morality

97. What makes something or someone "immoral"?
98. How do you know the difference between right and wrong?
99. What is the difference between morality and obedience?
100. Is it morally acceptable to take one human life to save two? If not, how many lives would have to be saved for the taking of one life to be morally acceptable?
101. If something is morally wrong, does that mean that it was always morally wrong and will always be morally wrong?
102. If something is morally wrong for one person, is it also morally wrong for everyone?
103. If something is morally wrong under one set of circumstances, is it morally wrong under all sets of circumstances?
104. Is it the act that is morally wrong, or the reasons behind the act, or does it not matter?
105. Is it morally wrong to commit suicide?
106. Is it morally wrong to end your life if you are terminally ill and suffering?
107. Should people be forced to share non-vital organs and fluids with people who need them to survive?
108. Should women be required by law (civil or moral) to see all pregnancies to term?
109. Is there anything morally wrong with using recreational drugs?
110. Is there anything morally wrong with getting drunk?
111. What is the difference between murder and "justified killing"?

112. If your $10 can feed a starving child in Africa for a month, is it right to spend your money on multiple pairs of shoes, movies, and other luxuries?
113. Is it morally wrong to be proud of your accomplishments?
114. Is it morally acceptable to kill animals and if so, why?
115. Why should humans' right to life and dignity be any different than a non-human life?
116. Would it be morally acceptable for an advanced life form to kill us for food or sport because they are exponentially more conscious and intelligent than we are?
117. Is valuing freedom more than safety right or wrong?
118. If you believe that God told you to kill someone, would you?
119. Is there always a right moral answer to every moral question (i.e. objectivism)?
120. Is doing the right thing for the wrong reason immoral?
121. Is doing the wrong thing for the right reason immoral?
122. Is there a difference between moral values and moral duties?
123. Is taking action to kill someone morally different than failing to act when you can easily and risk-free save someone's life?
124. Can something be moral in one culture, but immoral in another?
125. Is well-being what we ultimately consider the basis of morality? If so, how do we define well-being? Can people define it differently in that it changes what actions can be moral or immoral?
126. Well-being of whom? Self? Family? Country? Humanity? Other life forms? Nature? Is there an order of importance and if so, how do we determine the order?
127. Should we include the mental or psychological well-being of others?
128. Is it more moral to reduce suffering or increase happiness?
129. Is morality about the well-being of the maximum number of people or the maximum well-being of people?

130. To judge something as right or wrong, do we need a moral standard or just a moral comparison (like judging if a person is attractive or not)?
131. If there are objective moral facts, what are they?
132. Is lying ever acceptable?
133. Is retributive justice humane?

Mental Health

134. Are those who experience negative life events "traumatized," because we, as a society, coddle them too much?
135. How do we determine the difference between a psychological problem and an socially acceptable deviation from normal behavior?
136. Do people have any control over depression?
137. Why is homosexuality no longer classified as a mental disorder (it was until 1973)?
138. Is there really such thing as a sex addiction?

Politically Incorrect

139. Are some stereotypes accurate?
140. Are there professions that men are better at then women and vice versa?
141. Does the victim ever deserve some of the blame, if not all of it?
142. Is it wrong to prefer people who are most like us, even if they are only like us in skin color and gender?
143. Why is saying "people of color" appropriate but saying "colored people" is not?
144. Is life easier for good looking people than ugly people?
145. Are black people more likely to commit violent crime, or just more likely to get convicted?

146. Does being disgusted by watching two people of the same sex passionately kiss make someone homophobic?
147. What makes something or someone racist?
148. Can something be racist to one person and not to another? Who is right?
149. Is racism getting worse with increased sensitivity to racial differences?
150. What makes something or someone sexist?
151. Can something be sexist to one person and not to another? Who is right?
152. Is sexism getting worse with increased sensitivity to gender differences?
153. Are there certain groups that are more privileged than others?
154. Which group is more privileged, economically depressed white men or wealthy black men?
155. All things being equal, should minority job candidates be selected over non-minority job candidates?
156. Does the creation and promotion of minority groups help people to forget about differences or make these differences more salient?
157. Is it acceptable for white students to form a "white student alliance" in a predominately black school?
158. Are thoughts and prayers just a substitution for taking action that will actually help?
159. If you have a penis and a beard, but identify as a woman, is it still okay to use the women's bathroom?
160. Are fat people perfect the way they are?
161. Why is telling a woman she has beautiful eyes acceptable, but telling her she has beautiful breasts offensive?

On the Lighter Side

162. Do you think women who get their hair chopped off get compliments because it really looks good, or because saying nothing would simply be too awkward?

163. Does posting on social media about all the charity work and kind things you do for others make you an ass?

164. Does posting photos on social media of your expensive vacations that many of your friends could never afford make you an ass?

165. What if the author of this book is in cahoots with world leaders and Satan, and wrote this book as part of a master plan to destroy America and create godless heathens?

About The Author

For Dr. Bennett's complete bio, other books, online courses, and websites, please visit bobennett.com.